MS324 Waves, diffusion and variational principles

Block II

Random walks and diffusion

This publication forms part of an Open University course. Details of this and other Open University courses can be obtained from the Student Registration and Enquiry Service, The Open University, PO Box 197, Milton Keynes MK7 6BJ, United Kingdom: tel. +44 (0)845 300 6090, email general-enquiries@open.ac.uk

Alternatively, you may visit the Open University website at http://www.open.ac.uk where you can learn more about the wide range of courses and packs offered at all levels by The Open University.

To purchase a selection of Open University course materials visit http://www.ouw.co.uk, or contact Open University Worldwide, Walton Hall, Milton Keynes MK7 6AA, United Kingdom, for a brochure: tel. +44 (0)1908 858793, fax +44 (0)1908 858787, email ouw-customer-services@open.ac.uk

The Open University, Walton Hall, Milton Keynes, MK7 6AA.

First published 2005. Second edition 2010.

Copyright © 2005, 2010 The Open University

All rights reserved. No part of this publication may be reproduced, stored in a retrieval system, transmitted or utilised in any form or by any means, electronic, mechanical, photocopying, recording or otherwise, without written permission from the publisher or a licence from the Copyright Licensing Agency Ltd. Details of such licences (for reprographic reproduction) may be obtained from the Copyright Licensing Agency Ltd, Saffron House, 6–10 Kirby Street, London EC1N 8TS; website http://www.cla.co.uk.

Open University course materials may also be made available in electronic formats for use by students of the University. All rights, including copyright and related rights and database rights, in electronic course materials and their contents are owned by or licensed to The Open University, or otherwise used by The Open University as permitted by applicable law.

In using electronic course materials and their contents you agree that your use will be solely for the purposes of following an Open University course of study or otherwise as licensed by The Open University or its assigns.

Except as permitted above you undertake not to copy, store in any medium (including electronic storage or use in a website), distribute, transmit or retransmit, broadcast, modify or show in public such electronic materials in whole or in part without the prior written consent of The Open University or in accordance with the Copyright, Designs and Patents Act 1988.

Edited, designed and typeset by The Open University, using the Open University TeX System.

Printed in the United Kingdom by Cambrian Printer Limited, Aberystwyth.

ISBN 978 0 7492 5168 0

2.1

The paper used in this publication contains pulp sourced from forests independently certified to the Forest Stewardship Council (FSC) principles and criteria. Chain of custody certification allows the pulp from these forests to be tracked to the end use. (see www.fsc-uk.org).

Contents

INTRODUCTION AND OVERVIEW		**7**
CHAPTER 1 PROBABILITY AND STATISTICS		**13**
1.1	Probability	13
	1.1.1 Probability for discrete events	13
	1.1.2 Probabilities for combinations of events	15
	1.1.3 Probabilities for successive trials	16
	1.1.4 Continuous random variables	17
	1.1.5 Two or more random variables	20
1.2	Statistics	22
	1.2.1 Statistics of a single random variable	22
	1.2.2 Statistics involving two or more random variables	25
1.3	The normal distribution	28
	1.3.1 The error function	30
1.4	Outcomes	32
1.5	Further Exercises	33
	Solutions to Exercises in Chapter 1	35
CHAPTER 2 DISCRETE RANDOM FUNCTIONS AND RANDOM WALKS		**43**
2.1	Introduction	43
2.2	An elementary random function	44
	2.2.1 Defining the coin-tossing function	44
	2.2.2 Statistics of the coin-tossing function	45
2.3	Random walks	48
	2.3.1 Definition and mathematical description	48
	2.3.2 Some examples of random walks	50
2.4	Statistics of random walks	52
2.5	Probability distribution of a random walk	54
2.6	An approximate form for the probability distribution	57
2.7	Relationship with the diffusion equation	60
2.8	A random function with correlations	61
2.9	Summary and discussion	65
2.10	Outcomes	66
2.11	Further Exercises	66
	Solutions to Exercises in Chapter 2	68

CHAPTER 3   THE DIFFUSION EQUATION   75

- 3.1 Introduction — 75
- 3.2 Diffusion — 76
- 3.3 Concentration and flux density — 78
  - 3.3.1 Concentration — 78
  - 3.3.2 Molar concentration — 80
  - 3.3.3 Flux and flux density — 81
  - 3.3.4 Relating flux to the vector flux density — 83
  - 3.3.5 Defining flux density in one dimension — 85
  - 3.3.6 Summary of concentration and flux density — 86
- 3.4 The continuity equation — 87
  - 3.4.1 The continuity equation in one dimension — 87
  - 3.4.2 The continuity equation in three dimensions — 89
  - 3.4.3 Relation with Gauss's theorem — 91
- 3.5 The diffusion equation — 93
- 3.6 The heat equation — 95
- 3.7 A solution of the diffusion equation — 97
- 3.8 Summary and outcomes — 99
- 3.9 Further Exercises — 100
- 3.10 Appendix: observing diffusion in a simple experiment — 102
- Solutions to Exercises in Chapter 3 — 103

CHAPTER 4   SOLUTIONS OF THE DIFFUSION EQUATION   115

- 4.1 Introduction — 115
- 4.2 Diffusion in an infinite medium — 116
- 4.3 Solution in a finite medium in one dimension — 121
  - 4.3.1 Separation of variables — 122
  - 4.3.2 Identifying eigenfunctions — 123
  - 4.3.3 General solution — 123
  - 4.3.4 Orthogonality relation — 124
  - 4.3.5 Calculation of Fourier coefficients — 124
- 4.4 Orthogonality of eigenfunctions — 126
  - 4.4.1 Eigenfunctions and eigenvalues — 126
  - 4.4.2 Orthogonality of functions — 127
  - 4.4.3 Orthogonality relation — 128
- 4.5 Diffusion in a finite medium in two and three dimensions — 129
  - 4.5.1 The Helmholtz equation — 129
  - 4.5.2 A solution of the Helmholtz equation in two dimensions — 131
  - 4.5.3 Orthogonality and generalised Fourier series — 134
- 4.6 Semi-infinite domains: temperature waves — 137
- 4.7 Outcomes — 142
- 4.8 Further exercises — 142
- Solutions to Exercises in Chapter 4 — 144

CHAPTER 5 THE CENTRAL LIMIT THEOREM   153

5.1 Introduction   153
5.2 The central limit theorem   154
5.3 Distribution of sums of random variables   158
5.4 The Gaussian approximation to $[f(x)]^N$ (Optional)   163
5.5 Fourier transform of a probability density (Optional)   166
5.6 Summary   168
5.7 Outcomes   169
5.8 Further exercises   170
Solutions to Exercises in Chapter 5   171

CHAPTER 6 MICROSCOPIC DERIVATION OF THE DIFFUSION EQUATION   177

6.1 Introduction   177
6.2 Continuous random walks   179
    6.2.1 The continuous random walk in one dimension   179
    6.2.2 Random walks in two and three dimensions   181
6.3 From random walks to the diffusion equation   183
6.4 The Fokker–Planck equation   185
    6.4.1 Generalised diffusion processes   185
    6.4.2 Probability density for generalised diffusion   186
    6.4.3 Derivation of the Fokker–Planck equation   187
    6.4.4 The Fokker–Planck equation in three dimensions   190
6.5 Summary and outcomes   191
6.6 Further Exercises   191
Solutions to Exercises in Chapter 6   193

INDEX   197

Introduction and overview

This block is designed as an introduction to two related topics.

- *Random functions and random walks*

  The block introduces techniques used to model many of the types of random processes which occur in the natural world. It reviews the concepts of probability, random numbers and statistics, and their application to unpredictable phenomena. Sometimes it is desirable to understand how random quantities vary as a function of time (or some other variable), and for this reason it is useful to be able to define *random functions*, as well as random variables. We shall see how various types of random functions can be defined. One of these, the *random walk*, is a particularly important model, and is discussed in detail.

- *Diffusion and the heat equation*

  Diffusion is a process in which materials are mixed by random *microscopic* motion of atoms, without any large scale (*macroscopic*) motion (such as occurs when a liquid is stirred). Diffusion can be observed by dropping a pellet of dye into a glass bowl containing water which is left standing upon a cool floor. The dye dissolves, and over a period of several hours it gradually spreads away from the point where the dye pellet was standing. This block introduces the *diffusion equation*, which gives a *deterministic* (non-random) description of how the dye spreads. It also presents a derivation of the diffusion equation from the random microscopic motions of individual atoms, using the statistics of the random walk process.

  Processes described by the diffusion equation and closely related equations occur in many different areas of science. One application of the diffusion equation is to describe the flow of heat: it determines how the temperature of a body varies as a function of position and time. For this reason the diffusion equation is often called the *heat equation*, and many of the exercises are problems relating to the flow of heat.

Examples of random processes

This block is largely concerned with modelling random processes (also called *stochastic processes*). The techniques and concepts of probability and statistics are the tools required for understanding unpredictable events, so an elementary knowledge of these is required. The necessary concepts from the theory of probability and statistics are discussed in Chapter 1.

Before discussing this background material, it may be useful to describe some of the random processes to which these techniques might be applied. Figure 0.1 shows the number N of people standing in a queue, displayed as a function of time t. It is a mapping from the real line to the set of positive integers, which jumps by ± 1 at random times. The number of people standing in the queue at any one time is unpredictable. In order to describe this situation mathematically, it is necessary to develop models for random or *stochastic* functions, where the value of a function $N(t)$ is assigned by some random process.

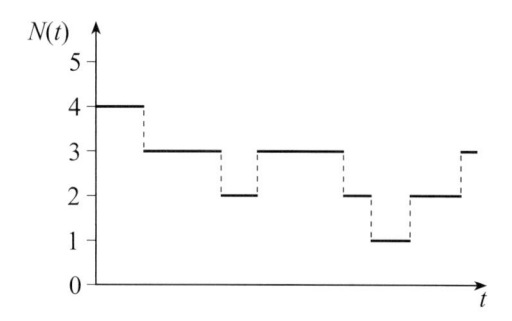

Figure 0.1 Number of people standing in a queue

There are many different types of stochastic processes, and some further examples are shown in the figures below. Figure 0.2 is a plot of the output voltage V of a Geiger counter, plotted as a function of time. The device produces a pulse whenever it detects a particle produced in a radioactive-decay process. The pulses all have the same shape, but the times at which they are produced are completely random.

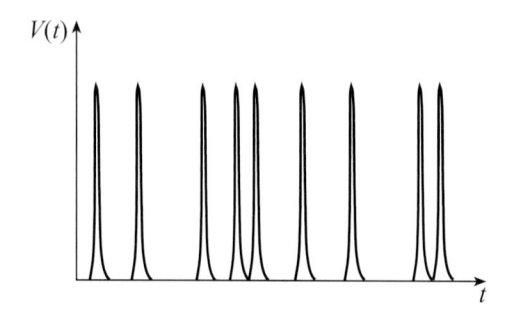

Figure 0.2 Output from a Geiger counter

Figure 0.3 shows the temperature recorded at midnight at a weather station in Cape Town on a succession of days starting at 4 January 2004. There are apparently unpredictable variations from the average. Even so, the successive numbers appear to be *correlated*, in that a day with a lower-than-average temperature is usually followed by another cooler-than-average day.

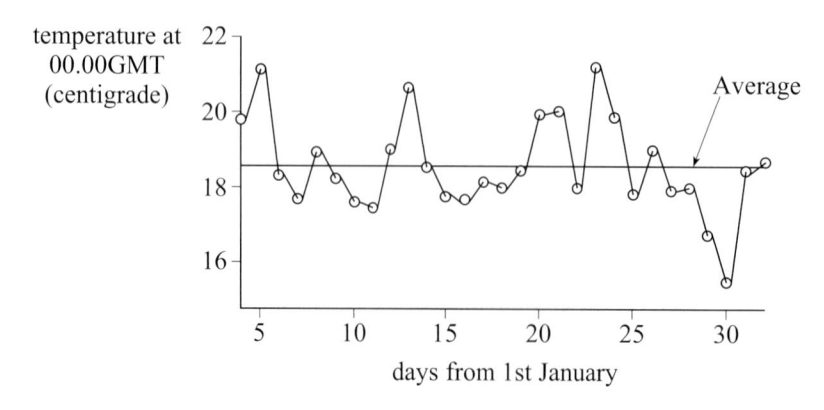

Figure 0.3 Temperature recorded at midnight in Cape Town on a series of January nights; the average value is shown for comparison. The values on successive days are correlated.

Figure 0.4 is a photograph of the surface of an ocean on a windy day. The height h of the surface (at a particular instant in time) is modelled by a continuous function of position (x, y) in the horizontal plane. An example of such a function is shown in Figure 0.5. Since $h(x, y)$ is a continuous function, knowing the height at one position gives an indication of what it might be at a nearby position. This is therefore another example of a random function with correlations.

Figure 0.4 Photograph of the mid-Atlantic Ocean (at sunset), showing a typical surface pattern

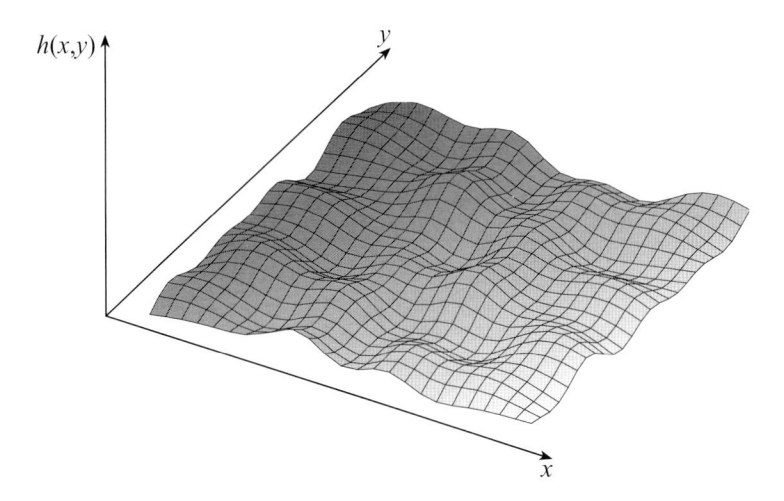

Figure 0.5 The height $h(x, y)$ of the ocean surface on a windy day is a random function of position

Figure 0.6 is a plot of one Cartesian component $x(t)$ of the position of a tiny dust particle moving in still water. The particle has a random motion (called *Brownian motion*) due to the fact that it is being jostled by movement of the water molecules. Like Figure 0.5, this is a mapping from one continuous variable to another, but it is very different in character. First, this is not a smoothly varying function. (The function is continuous, but its derivative is not well defined at any point, that is, it is nowhere differentiable.) Also, the magnitude of $x(t)$ is tending to increase, whereas the typical value of the function plotted in the previous examples does not change as time progresses. This example is modelled by a type of random function called a *random walk*, which is discussed in some detail in this block.

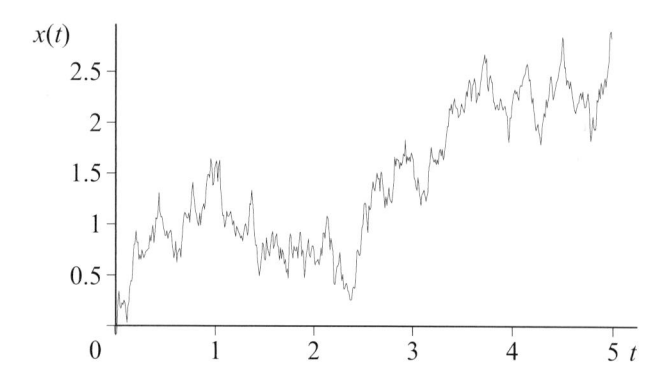

Figure 0.6 Brownian motion: one Cartesian component of the displacement of a tiny dust particle in still water as a function of time

Our final example, shown in Figure 0.7, is a plot of the value of a market price as a function of time, namely the closing price of gold (in US dollars per ounce) on a sequence of 996 trading days, from 1 January 1998 to 27 December 2001. Its character is clearly similar to Figure 0.6, in that the fluctuations are also very erratic, and there is no reason for the function to remain close to any particular value. The time-dependence of prices of commodities, shares and other traded assets is often modelled by a type of random walk.

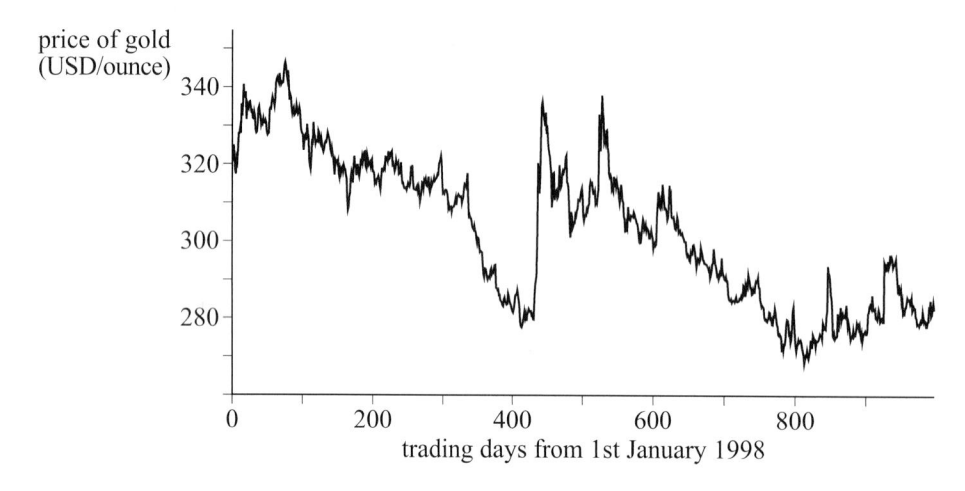

Figure 0.7 The price of gold (in US dollars per ounce) over the four years from January 1998 to December 2001

Many other examples of random processes can be observed in everyday life. Of course, it is (by definition) impossible to predict the precise course of these random processes. But it is an important task for applied mathematicians to be able to describe them as fully as possible. This requires the use of statistical techniques, in which these processes are modelled by *random functions*, which are introduced in Chapter 2.

The relation between random and deterministic processes

There is one type of situation in which random motions do lead to predictable results. This happens when we observe the motion of a large number of randomly moving particles. When the numbers of particles are very large, the number of particles in a given region of space, N say, can be predicted very accurately (in the sense that the ratio of the error δN to N is small, although δN itself may be very large). In these situations, accurate predictions are possible even though the motions of individual particles are random. The block discusses the derivation and solution of the *diffusion equation*, a partial differential equation which describes the motion of large numbers of randomly moving particles.

Chapter 3 introduces the diffusion equation, describing diffusion as a macroscopic phenomenon, giving a derivation based upon a plausible assumption, which is sometimes known as *Fick's law*. Chapter 4 explores the solutions of the diffusion equation; here many of the techniques (separation of variables, Fourier analysis) have already been encountered in finding solutions of the wave equation.

Later chapters consider the relationship between the deterministic diffusion equation and the random microscopic motions which give rise to diffusion. Chapter 5 discusses the *central limit theorem*, a very general and powerful concept in probability theory, which can be used to explain the connection between random walks and diffusion in a restricted case. Chapter 6 discusses the derivation of the diffusion equation from a microscopic theory, initially using the central limit theorem, and then by giving a derivation of a rather general form of the diffusion equation, called the Fokker–Planck equation.

> It may be useful to include a few words about closely related areas which will not be treated here.
>
> A vast range of phenomena can be described by random processes, and we make no attempt to give anything like a complete or systematic treatment. Numerous other examples can be found by consulting books on *stochastic processes*. Our discussion of random functions will concentrate on the random walk, because of its close connection with the diffusion equation.
>
> There is also an important area of theoretical physics known as *statistical mechanics*, which combines statistical and mechanical principles to define quantities such as temperature, and which leads to methods for calculating properties of materials in *thermal equilibrium*. This too is outside the scope of this course. Although the discussion of the microscopic mechanism of diffusion involves a combination of statistical and dynamical ideas, it is distinct from statistical mechanics in that this course does not require the mechanical definition of temperature.

CHAPTER 1
Probability and statistics

This chapter discusses the concepts from probability and statistics which you will need for the remainder of this block. Many of the ideas may be familiar from earlier courses, but the notation, terminology and definitions may be different.

1.1 Probability

1.1.1 Probability for discrete events

If you perform some procedure where the set of possible outcomes is known, but the actual outcome is uncertain, it is useful to describe each possible outcome in terms of its probability. The simplest example is tossing a coin, where there are two possible outcomes, namely the coin landing heads up or tails up. This is a commonly used method to make an unbiased random decision between two possibilities. Most people would be comfortable with the statement that both outcomes are equally likely. The equivalent mathematical statement is that both outcomes have probability equal to one half. We start by reviewing how probabilities are defined.

Suppose that we undertake a particular 'trial' or experiment in which there are n possible distinct outcomes:

$$\text{outcome 1, outcome 2, } \ldots, \text{ outcome } n. \tag{1.1}$$

We perform the experiment N times, and count how many times each outcome occurs, say

$$N_1, N_2, \ldots, N_n, \tag{1.2}$$

where N_i is the number of times outcome i is observed; N_i is sometimes referred to as the frequency of outcome i. We now examine the values of N_i/N as N increases. Provided that the circumstances of the trial do not change, we expect that the values of N_i/N will approach a limit as N increases. The probability P_i of the occurrence of outcome i is defined as

$$P_i = \lim_{N \to \infty} \frac{N_i}{N}. \tag{1.3}$$

The sum of the number of times N_i that each distinct outcome occurs is equal to N, i.e. $\sum_{i=1}^{n} N_i = N$. The sum of the probabilities of all possible outcomes is therefore

$$\sum_{i=1}^{n} P_i = \sum_{i=1}^{n} \lim_{N \to \infty} \frac{N_i}{N} = \lim_{N \to \infty} \frac{1}{N} \sum_{i=1}^{n} N_i = \lim_{N \to \infty} 1, \tag{1.4}$$

Throughout this block we shall use the terms 'outcome' and 'event' interchangeably.

so
$$\sum_{i=1}^{n} P_i = 1. \tag{1.5}$$

Sometimes probabilities are known on theoretical grounds. For example, when a coin is tossed, there are $n = 2$ outcomes, and we expect that the coin is as likely to fall 'tails' as it is to fall 'heads' (assuming that the coin is fair). One therefore expects that the probability of falling heads, P_h, and the probability of falling tails, P_t, are equal (and equal to one half, because equation (1.5) gives $P_t + P_h = 1$). Similarly, a six-sided die is expected to have an equal probability of falling with any face upwards, i.e. $P_i = \frac{1}{6}$ for $i = 1, \ldots, 6$. Any trial where nothing favours any one of n possible outcomes has probability $P_i = 1/n$ for any outcome.

Exercise 1.1

Try tossing a coin N times, and record the number of heads N_h. As you increase N, you will find that N_h/N approaches one half, slowly and erratically. Do not let N be too large; about 40 should suffice for this experiment.

In other circumstances, probabilities are not known exactly, and must be determined *empirically*, by repeated trials. An example would be a surgical procedure where the outcome is either success or failure. If an operation is performed 943 times, and the outcome is successful 455 times, the probability of success may be presumed to be close to $455/943 \simeq 0.48$. There are interesting problems associated with deciding how accurate this estimate is. For example, could you reliably conclude that a variation of the operating procedure, which had been successful on 7 out of 10 occasions, should be adopted in preference to the original? Such questions are treated in many texts on statistics, but they will be ignored in this course. Here we shall be concerned with situations in which the probabilities for the whole set of possible outcomes can be deduced from theoretical arguments. However, the following example illustrates an interesting use of empirical estimates of probabilities.

Example 1.1

Let P_e be the probability that a letter chosen at random on a page of English text is the letter e. Estimate P_e by counting letters in the first sentence of the preceding paragraph (that is, the sentence starting 'In other circumstances ...'), then in the whole of that paragraph.

Solution

In the first sentence there are 97 letters, of which 13 are the letter e, giving the estimate $P_e \simeq 13/97 = 0.134$. In the whole paragraph there are 705 letters, of which 98 are the letter e, giving $P_e \simeq 98/705 = 0.139$. The frequency of the letter e in longer texts is close to this value; for example, in the book of Psalms, in the King James Bible, it has been quoted as $P_e \simeq 0.13$. Incidentally, the use of estimates of letter frequencies has often been used to decipher simple substitutional codes; an amusing example of this can be found in one of Arthur Conan Doyle's Sherlock Holmes stories, 'The adventure of the dancing men'. ■

1.1.2 Probabilities for combinations of events

Sometimes we are interested in outcomes that are combinations of different *elementary outcomes* for which the probabilities are already known. (Elementary outcomes are the simplest possible outcomes in terms of which all other more complicated outcomes can be expressed.) For example, what is the probability of throwing a number less than 3 (either 1 or 2) with a six-sided die? The number of times this happens is $N_{1\,\text{or}\,2} = N_1 + N_2$, where N_1 and N_2 are the numbers of trials where 1 and 2 are thrown, respectively. It follows from equation (1.3) that the probability of throwing 1 or 2 is $P_{1\,\text{or}\,2} = P_1 + P_2$. (We know that $P_1 = \frac{1}{6}$ and $P_2 = \frac{1}{6}$ are the probabilities of throwing 1 and 2, respectively, so $P_{1\,\text{or}\,2} = \frac{1}{3}$.) In general, you can add the probabilities of outcomes which are *mutually exclusive events* to give the probability of the *combined* event: the probability of observing outcome A or outcome B is

$$P_{A\,\text{or}\,B} = P_A + P_B. \tag{1.6}$$

As an example of two events which are not mutually exclusive, a person can have both blonde hair and blue eyes. In cases like this, the generalisation of equation (1.6) is

$$P_{A\,\text{or}\,B} = P_A + P_B - P_{A\,\text{and}\,B}, \tag{1.7}$$

where $P_{A\,\text{and}\,B}$ is the probability of outcomes A and B occurring in the same trial. (Subtracting the term $P_{A\,\text{and}\,B}$ avoids double-counting the cases where events A and B both happen.) This more general version of equation (1.6) will not be required in the remainder of the course.

Exercise 1.2

What is the probability of drawing a picture card (King, Queen or Jack) from a shuffled deck of 52 playing cards?

Exercise 1.3

The table below lists the number of deaths recorded in England in a given year, for people in given age ranges. Use this table to estimate the probability that someone will die between the ages of 25 and 64. What assumption have you used?

Table 1.1

Age range	Number of deaths
0–4	6 000
5–14	1 200
15–24	4 400
25–44	14 600
45–64	88 200
65–74	141 300
75+	319 500

1.1.3 Probabilities for successive trials

In some cases we might need probabilities for events at successive trials. If the trials are independent, meaning that the outcome of one trial does not influence another, the probabilities are multiplied. For example, the probability of drawing an ace from a deck of cards is $4/52 = 1/13$, so the probability of drawing an ace on two successive independent trials, with the card that was drawn replaced and the deck shuffled, is $\frac{1}{13} \times \frac{1}{13} = 1/169$. In general, if a trial is repeated, then the probability of obtaining result 'a' on the first trial and 'b' on the second trial is

$$P_{a,b} = P_a P_b, \tag{1.8}$$

with $P_{a,b}$ denoting the probability of the events occurring in succession. This applies only if the trials are independent. In the above example, if the first card chosen is not returned to the deck, then the probability of drawing a second ace from the deck, having found one on the first draw, is $3/51$, and the probability of two aces in succession is $\frac{4}{52} \times \frac{3}{51} = 1/221$.

When successive events are not independent, the probability of event 'b' following event 'a' is written

$$P_{a,b} = P_a P_{b\,\text{given}\,a}, \tag{1.9}$$

where $P_{b\,\text{given}\,a}$ is the probability that event 'b' happens on the second trial if event 'a' has occurred on the first trial. In the preceding example, events 'a' and 'b' are both drawing an ace from the deck, but the successive trial is such that the second card is drawn while *not returning* the first drawn card to the deck, so we have $P_a = \frac{1}{13}$ and $P_{b\,\text{given}\,a} = \frac{3}{51}$.

In the remainder of this course we shall always be dealing with situations where events are independent, so that $P_{b\,\text{given}\,a} = P_b$, and equation (1.8) can be used. This would be the case when events 'a' and 'b' are drawing aces from the deck but drawing the second card only after *returning* the first drawn card to the deck and reshuffling.

Exercise 1.4

When throwing two six-sided dice at the same time, what is the probability that no number thrown on either die is higher than two?

First derive the result from equation (1.8), then check it by counting all of the possible pairs of throws with neither number higher than two.

Exercise 1.5

If a sport can be practised only on approximately four days out of ten due to weather limitations, and an enthusiast has time available only at weekends, what is the probability that she can practise the sport on any given day? Approximately how many suitable days would be available each year?

Exercise 1.6

Show that when two six-sided dice are thrown, the probability that the sum of the scores is equal to seven is $\frac{1}{6}$.

1.1.4 Continuous random variables

We can generalise those concepts developed for discrete events to the case of trials having a continuous range of possible outcomes, such as when a measurement of a quantity yields a real number x. An example of such a trial would be the measurement of the height in cm of a person chosen at random from a large population. (We are considering an idealised situation where a person's height can be measured with such great precision that it may be regarded as a continuous variable.) When the set of possible values of x is continuous, it is not useful to ask for the probability that x takes a given value, such as x_0, because (except in exceptional circumstances) the number will never be precisely equal to x_0. We can ask, instead: 'What is the probability that x lies in an interval between x_0 and $x_0 + \delta x$?' For example, it is meaningful to ask the question 'What is the probability that an adult will have a height between 190 cm and 191 cm?', but it is not useful to ask 'What is the probability that an adult will have a height of *exactly* 190 cm?', since this probability would be vanishingly small if the height is measured to arbitrary accuracy.

The probability that the height of an individual selected from a very large specified population is between x_1 and x_2 will be a function of x_1 and x_2, and will be written $P(x_1, x_2)$. (We shall assume throughout that $x_2 > x_1$.) Now, if $\delta x = x_2 - x_1$ is small, it is natural to expect that this probability will be proportional to δx. (In the example given above, you might reasonably expect that the probability for a person's height to be between 190 cm and 192 cm is roughly double the probability that it is between 190 cm and 191 cm.) A more precise statement is that one expects that

$$\rho(x) = \lim_{\delta x \to 0} \frac{P(x, x + \delta x)}{\delta x} \qquad (1.10)$$

exists for a sufficiently well-behaved $P(x_1, x_2)$ (i.e. partial derivatives of $P(x_1, x_2)$ exist at $x_1 = x_2 = x$). The function $\rho(x)$ is called the *probability density* of the *continuous random variable* x. Note that this function cannot take negative values, because the probability is never negative.

We shall often find it convenient to discuss the probability δP that x lies in a small interval. We shall refer to this as the *element of probability*: δP is the probability that the random variable lies in the interval $[x, x + \delta x]$ with a *small* width δx. We have

$$\delta P = P(x, x + \delta x) = \rho(x)\,\delta x + O(\delta x^2) \qquad (1.11)$$

(where δx is assumed to be positive). The approximation $\delta P \simeq \rho(x)\,\delta x$ is valid for sufficiently small values of δx (provided that $P(x_1, x_2)$ is differentiable). The probability $P(x_1, x_2)$ may be expressed as an integral of the probability density function:

$$P(x_1, x_2) = \int_{x_1}^{x_2} dx\, \rho(x). \qquad (1.12)$$

It is worth emphasising at this point that equation (1.12) can serve as a more general definition of the probability density function; it does not require the assumption that $P(x_1, x_2)$ be differentiable, as was needed for equations (1.10) and (1.11). Of course, for a sufficiently well-behaved $\rho(x)$, equation (1.11) follows directly and simply from equation (1.12), and Example 1.2, below, will show that equation (1.10) can follow directly from equation (1.12). For these reasons, equation (1.12) is usually taken to be the preferred definition of $\rho(x)$.

Because the continuous random variable x has to take some value between $-\infty$ and $+\infty$, we have

$$P(-\infty, \infty) = \lim_{x_1 \to -\infty} \lim_{x_2 \to +\infty} P(x_1, x_2) = 1. \quad (1.13)$$

It follows from equation (1.12) that the probability density satisfies $\rho(x) \geq 0$ and

$$\int_{-\infty}^{\infty} dx\, \rho(x) = 1. \quad (1.14)$$

A function must satisfy the *normalisation condition* (1.14) if it is to be a valid probability density. A function which satisfies condition (1.14) is said to be *normalised*.

In many cases it is known that the random variable x lies in a range between some numbers $x_{\min}$ and $x_{\max}$, so that $\rho(x)$ vanishes for $x < x_{\min}$ and for $x > x_{\max}$. It follows that in this case equation (1.14) becomes

$$\int_{x_{\min}}^{x_{\max}} dx\, \rho(x) = 1. \quad (1.15)$$

Example 1.2

Show that equation (1.12) implies equation (1.10). [Hint: Consider the partial derivative of equation (1.12) with respect to x_2, and assume it exists.]

Solution

From the definition of the partial derivative, equation (1.10) can be rewritten as

$$\rho(x_1) = \left.\frac{\partial P}{\partial x_2}(x_1, x_2)\right|_{x_2 = x_1} \quad (1')$$

(that is, we take the partial derivative with respect to x_2, then set $x_2 = x_1$). Differentiating equation (1.12) gives

$$\frac{\partial P}{\partial x_2}(x_1, x_2) = \frac{\partial}{\partial x_2} \int_{x_1}^{x_2} dx\, \rho(x) = \rho(x_2); \quad (2')$$

then setting $x_2 = x_1$ gives the same result as equation $(1')$. ■

It can often be useful to think in terms of a graph of the probability density $\rho(x)$. Equation (1.12) shows that the probability that the random variable takes a value between x_1 and x_2 is the area under the curve for $\rho(x)$ between x_1 and x_2 (see Figure 1.1).

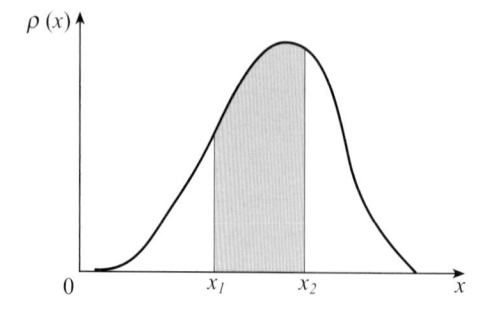

Figure 1.1 The probability that a random variable x takes a value between x_1 and x_2 is equal to the integral of the probability density $\rho(x)$ between these two limits (that is, the area of the shaded region under the curve)

One very simple probability density is the *uniform distribution*, where the probability density $\rho(x)$ for the random variable x is independent of x for all

x within some interval, and zero elsewhere. A uniform distribution for a random variable which lies between x_1 and x_2 (where $x_2 > x_1$) has probability density

$$\rho(x) = \begin{cases} \text{constant}, & x_1 \leq x \leq x_2, \\ 0, & \text{otherwise}. \end{cases} \tag{1.16}$$

Exercise 1.7

By considering the normalisation of the probability density equation (1.16), show that the value of the constant is $1/(x_2 - x_1)$.

Exercise 1.8

The pilot of an aircraft is expected to check the tyre pressures before each flight. The valve on one of the wheels is hidden behind other structures when it is rotated less than 20° either side of the vertical line from the axle, in which case the aircraft will have to be moved before the tyre pressure can be checked (see Figure 1.2).

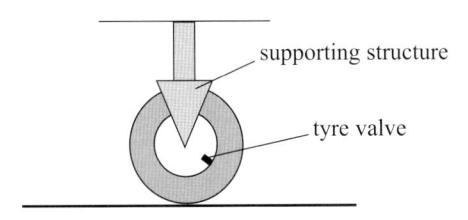

Figure 1.2 What is the probability that the tyre valve will be hidden when the aircraft is parked?

What is the probability density for the angle of the tyre valve relative to the vertical? What is the probability that the aircraft will have to be moved to allow access to the valve?

Exercise 1.9

(a) The length of time that you have to wait for a bus is apparently random. The time t may be regarded as a random variable with a probability density $\rho(t)$. Some plausible assumptions indicate that for $t > 0$, ρ is an exponential function $\rho(t) = A \exp(-t/t_0)$, where t_0 is a typical waiting time. Show that the probability density is normalised if $A = 1/t_0$. These assumptions are discussed in Exercise 1.32 at the end of this chapter.

(b) A regular passenger (with an understanding of probability) knows that $t_0 = 15$ minutes. He decides that if his bus takes longer than 45 minutes to appear, then the drivers have probably gone on strike, and he walks home. What is the probability that he will walk home if the buses are running normally?

1.1.5 Two or more random variables

It is often necessary to consider situations where two or more random variables are measured or observed. The probability density for measurement of two or more random variables is defined in a similar way to that of a single random variable. For example, the height h and weight w of a person drawn at random from a large population can be regarded as two random variables. The probability δP that the height of a person lies between h and $h + \delta h$, and his or her weight lies between w and $w + \delta w$, is expected to be proportional to both of the small increments δh and δw. It may be written as a generalisation of equation (1.11):

$$\delta P \simeq \rho(h, w)\, \delta h\, \delta w, \tag{1.17}$$

where $\rho(h, w)$ is the *joint probability density* for height and weight.

In general, given two random variables x and y, the element of probability δP that x lies in the interval $[x, x + \delta x]$ and y lies in the interval $[y, y + \delta y]$ (where δx and δy are small) is written

$$\delta P \simeq \rho(x, y)\, \delta x\, \delta y. \tag{1.18}$$

The quantity $\rho(x, y)$ is called the *joint probability density function for x and y*.

Equation (1.12) connects the probability density with a probability for a single random variable. Let us consider the analogous relation for two random variables, which gives the probability $P_\mathcal{A}$ that the point (x, y) lies inside a region $\mathcal{A}$ in the (x, y)-plane (see Figure 1.3).

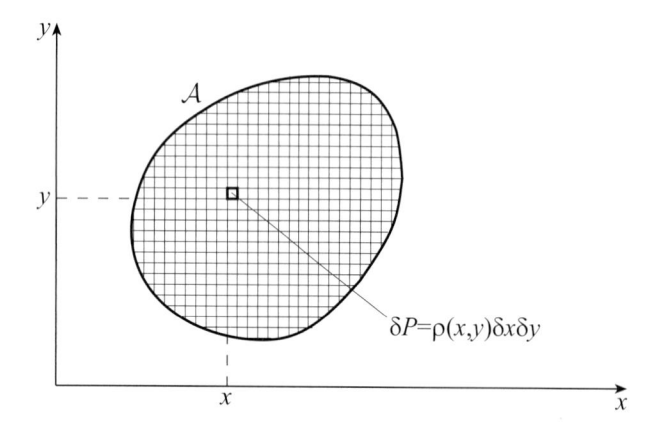

Figure 1.3 The probability that a pair of random variables takes values lying inside a small rectangle of size $\delta x \times \delta y$ at (x, y) is $\rho(x, y)\, \delta x\, \delta y$. The probability that the pair of random variables (x, y) lies in a region $\mathcal{A}$ is the integral of ρ over the region $\mathcal{A}$.

Summing the contributions from every element, and taking the limit as $\delta x, \delta y \to 0$, we see that $P_\mathcal{A}$ is expressed as an area integral of the density $\rho(x, y)$ over the region $\mathcal{A}$:

$$P_\mathcal{A} = \iint_\mathcal{A} dx\, dy\, \rho(x, y). \tag{1.19}$$

Again, just as equation (1.12) can be regarded as the more fundamental definition of the probability density function $\rho(x)$, equation (1.19) is to be viewed as the preferred definition of the joint probability density $\rho(x, y)$.

If the joint probability density factorises such that

$$\rho(x, y) = \rho_1(x)\, \rho_2(y), \tag{1.20}$$

where $\rho_1(x)$ and $\rho_2(y)$ are the probability densities for single observations of each of the variables, then these random variables are said to be *independent*. This expression is analogous to equation (1.8) for the case of discrete probability distributions. The term 'independent' is being used in the same sense as for discrete random variables, because equation (1.20) implies that determining x has no influence on a measurement of y (and vice versa).

The following is an example of a joint probability density for two independent random variables. The length of time that someone has to wait for their bus in the morning is t_1. This is a random variable, with probability density $\rho_1(t_1)$. The waiting time for the bus for the evening return journey is t_2, with probability density $\rho_2(t_2)$. The probability that the bus arrives after a waiting time between t_1 and $t_1 + \delta t_1$ is $\delta P_1 = \rho_1(t_1)\,\delta t_1$. It is very reasonable to assume that the waiting time in the morning has no relation to that in the evening, in that whatever the value of t_1, the probability that the second wait is between t_2 and $t_2 + \delta t_2$ is $\delta P_2 = \rho_2(t_2)\,\delta t_2$. Equation (1.8) implies that the probability that the waiting time in the morning is between t_1 and $t_1 + \delta t_1$, while that in the evening is between t_2 and $t_2 + \delta t_2$, is $\delta P = \rho_1(t_1)\,\rho_2(t_2)\,\delta t_1\,\delta t_2$. It is also, by definition, $\delta P = \rho(t_1, t_2)\,\delta t_1\,\delta t_2$, where $\rho(t_1, t_2)$ is the joint probability density, so

$$\rho(t_1, t_2) = \rho_1(t_1)\,\rho_2(t_2), \tag{1.21}$$

in agreement with equation (1.20).

Exercise 1.10

In the situation described above, what is the joint probability density for the waiting times if t_1 and t_2 both have the exponential probability density considered in Exercise 1.9, with $t_0 = 15$ minutes? What is the probability of waiting less than 15 minutes in the morning and less than 15 minutes in the evening?

Exercise 1.11 *Harder exercise*

In the situation described in the previous exercise, what is the probability of the total waiting time $t_1 + t_2$ being less than 15 minutes?

Exercise 1.12

What is the condition for a joint probability density to be normalised? What other condition must a function satisfy if it is to be a valid joint probability density?

Exercise 1.13 *Harder exercise*

Two random variables, x and y, have joint probability density $\rho(x, y)$. Show that the probability density for the random variable x, which is here denoted by $\rho_1(x)$, is obtained from $\rho(x, y)$ by integrating over y:

$$\rho_1(x) = \int_{-\infty}^{\infty} dy\, \rho(x, y). \tag{1.22}$$

Verify that this integral gives the correct result when x and y are independent variables.

In most cases where we consider probabilities for pairs of random numbers, we shall be considering variables with a continuous distribution, described by a joint probability density. Occasionally we shall consider pairs of random variables which take discrete values. In this case we can define a *joint probability* for the two variables. Consider the case where two random variables X and Y take, respectively, n and m possible discrete values, X_i and Y_j, labelled by the integers $i = 1, \ldots, n$ and $j = 1, \ldots, m$, respectively. The joint probability $P_{i,j}$ is the probability that X takes the value X_i and Y takes the value Y_j. The results considered above all generalise in a natural way to the discrete case. For example, the normalisation condition becomes

$$\sum_{i=1}^{n}\sum_{j=1}^{m} P_{i,j} = 1, \qquad (1.23)$$

which is analogous to the solution of Exercise 1.12. Also, the formula to obtain the probability P_i that X takes the value X_i is

$$P_i = \sum_{j=1}^{m} P_{i,j}, \qquad (1.24)$$

which is analogous to equation (1.22).

1.2 Statistics

1.2.1 Statistics of a single random variable

Probabilities and probability densities give the fullest possible description of random variables. An accurate determination of the probability density function of a random variable from repeated trials would require an enormous number of observations. For this reason, it is customary to describe a random variable by means of *statistics*, which are single numbers, such as average values, describing some aspect of the range of values taken by the random variable. For many purposes, probability densities contain too much information to be directly useful, whereas statistics give a concise summary of what should be expected.

> The term *statistic* is popularly used for any piece of information in numerical form, for example the number of tons of steel produced last month in Germany. In our terminology such a number would be not a statistic, but a single observation. However, the average monthly steel production over a three-year period would be a statistic, because it is a single number describing a set of observations.

The simplest (and best known) statistic is the average. Given a set of N numbers $x_1, x_2, \ldots, x_i, \ldots, x_N$, the average is

$$x_{\mathrm{av}} = \frac{1}{N}\sum_{i=1}^{N} x_i. \qquad (1.25)$$

The average will depend upon the particular set of observations, but in the case where the number of observations N approaches infinity, we expect that the average will approach a limit. We now investigate the limit as $N \to \infty$, considering first the case where each observation x_i can take only one of n discrete values, with the possible values of x_i being labelled X_j, $j = 1, \ldots, n$, and with each value occurring N_j times (with $N = \sum_{j=1}^{n} N_j$). Then the average may be written as

$$x_{\mathrm{av}} = \sum_{j=1}^{n} \frac{N_j}{N} X_j. \qquad (1.26)$$

1.2 Statistics

Example 1.3

Throwing a six-sided die five times gave the results $x_1 = 3$, $x_2 = 6$, $x_3 = 3$, $x_4 = 6$, $x_5 = 4$, where x_i is the number obtained on throw i. Throwing the same die 25 times gave the sequence

$$4, 6, 5, 3, 6, 3, 2, 2, 2, 4, 4, 3, 3, 2, 1, 4, 4, 6, 1, 1, 1, 2, 4, 2, 1.$$

Calculate the average of each sequence.

These sequences take a discrete set of values, with $n = 6$ possibilities. What are the values of X_j? And what are the values of N_j for the second sequence? Verify that for the second sequence, equation (1.26) gives the same answer as the average determined in the first part of this question.

Solution

The sum of the first sequence is 22, so the average is $22/5 = 4.4$. The sum of the second sequence is 76, so the average is $76/25 = 3.04$.

The values of X_j are $X_1 = 1, \ldots, X_j = j, \ldots, X_6 = 6$, and their frequencies in the set of 25 throws are $N_1 = 5$, $N_2 = 6$, $N_3 = 4$, $N_4 = 6$, $N_5 = 1$, $N_6 = 3$, so

$$\sum_{j=1}^{n} X_j N_j = (1 \times 5) + (2 \times 6) + (3 \times 4) + (4 \times 6) + (5 \times 1) + (6 \times 3)$$
$$= 76,$$

which gives the same value for the average. ∎

In the limit as $N \to \infty$, equation (1.26) may be written in terms of probabilities:

$$\lim_{N \to \infty} x_{\text{av}} = \sum_{j=1}^{n} X_j \lim_{N \to \infty} \frac{N_j}{N} = \sum_{j=1}^{n} P_j X_j. \quad (1.27)$$

Except in some special cases where n is infinite, the average approaches a definite limit as $N \to \infty$. This limit is known as the *mean value* or *expectation value*, and is given the symbol $\langle x \rangle$. Angular brackets will be used to denote taking the limit of the average of the quantity inside the brackets, when the number of observations N approaches infinity. The mean value is given by the expression

$$\langle x \rangle = \sum_{j=1}^{n} P_j X_j. \quad (1.28)$$

Exercise 1.14

For the situation considered in Example 1.3 (throwing a six-sided die), what should be the limit of the average value as the number of throws is increased?

This formula for the mean generalises to the case of a continuous random variable with probability density $\rho(x)$. The generalisation can be derived by dividing the set of possible values of x into small intervals of length δx. Let us estimate the number of times, N_j, that the variable x lies between $x_j = j \, \delta x$ and $x_{j+1} = x_j + \delta x$ within a total number of N observations. This is approximately equal to N times the probability that x lies in this interval, that is, $N_j \simeq N\rho(x_j)\delta x$. The mean value is

$$\begin{aligned}\langle x \rangle &= \lim_{N \to \infty} x_{\text{av}} \\ &\simeq \lim_{N \to \infty} \sum_j \frac{N_j}{N} x_j \\ &\simeq \sum_j \delta x\, \rho(x_j)\, x_j.\end{aligned} \qquad (1.29)$$

Taking the limit as $\delta x \to 0$, the sum becomes an integral:

$$\langle x \rangle = \int_{-\infty}^{\infty} dx\, \rho(x)\, x. \qquad (1.30)$$

At this point it may be useful to review the discussion of definite integrals in Block 0.

Exercise 1.15

Calculate the mean value of a random variable x which has a probability density which is uniform on the interval $[x_1, x_2]$.

Exercise 1.16

Calculate the mean of the exponential distribution introduced in Exercise 1.9, namely $\rho(x) = \exp(-x/x_0)/x_0$ for $x \geq 0$, and $\rho(x) = 0$ for $x < 0$.

The mean value is not the only statistic which is of interest. The *moments* M_k of a random variable x are defined to be the mean values of x^k (where k is a non-negative integer): in the case where x has a continuous distribution, the kth moment is

$$M_k = \langle x^k \rangle = \int_{-\infty}^{\infty} dx\, x^k \rho(x). \qquad (1.31)$$

Clearly, M_1 is the mean, and (recalling the normalisation condition (1.14)) $M_0 = 1$.

In some cases the integral (1.31) may diverge, and one may encounter a probability density function $\rho(x)$ for which only one or two moments exist. (Clearly, $M_0 = 1$ has to exist for $\rho(x)$ to be a valid probability density function.)

Exercise 1.17

Consider a uniformly distributed random variable x on the interval $[0, 1]$. Determine the moments M_k.

Exercise 1.18

Sometimes we may consider moments $\langle x^k \rangle$ in the case where x has a discrete distribution. Write down the expression for the moment $\langle x^k \rangle$ when the random variable takes n discrete values X_j, with probabilities P_j, $j = 1, \ldots, n$.

Determine $\langle x^2 \rangle$ where x is the number uppermost on throwing a die.

Having defined the mean value of a random variable x, it is useful to give a statistic describing how wide is the variation of its values. An appropriate statistic is called the *standard deviation*, denoted by σ. The square of σ is called the *variance*, written $\text{Var}(x)$, and it is defined as the mean value of the square of the difference Δx between the measured value and the mean value. Mathematically, these quantities are defined by the relations

$$\sigma^2 = \text{Var}(x) = \langle \Delta x^2 \rangle, \quad \Delta x = x - \langle x \rangle. \qquad (1.32)$$

1.2 Statistics

The variance can be expressed in a form which is more convenient for calculation, in terms of the second moment. Expanding the square in equation (1.32) and using equation (1.31) gives

$$\text{Var}(x) = \int_{-\infty}^{\infty} dx\, [x^2 - 2\langle x \rangle x + \langle x \rangle^2]\, \rho(x)$$
$$= \int_{-\infty}^{\infty} dx\, x^2 \rho(x) - 2\langle x \rangle \int_{-\infty}^{\infty} dx\, x\, \rho(x) + \langle x \rangle^2 \int_{-\infty}^{\infty} dx\, \rho(x). \quad (1.33)$$

The integrals in this expression are, respectively, the moments M_2, $M_1 = \langle x \rangle$ and $M_0 = 1$, so

$$\text{Var}(x) = \langle x^2 \rangle - \langle x \rangle^2 = M_2 - M_1^2. \quad (1.34)$$

This expression usually gives the simplest route to calculating the variance.

Exercise 1.19

Consider a uniformly distributed random variable x on the interval $[0, 1]$. Show that the variance of x is $\text{Var}(x) = \frac{1}{12}$.

1.2.2 Statistics involving two or more random variables

When we have more than one random variable, it becomes harder to gain accurate information about probability density functions. Also, even if the probability density function is known, it is harder to interpret functions of several variables. Using statistics to convey information about random variables becomes especially useful when dealing with two or more random variables.

Exercise 1.20

Two random variables have a joint probability density $\rho(x, y)$. Using the result of Exercise 1.13, show that the mean value of x is

$$\langle x \rangle = \int_{-\infty}^{\infty} dx \int_{-\infty}^{\infty} dy\, x\, \rho(x, y). \quad (1.35)$$

Given a joint probability density $\rho(x, y)$, the most general statistic we consider is $\langle f(x, y) \rangle$, where $f(x, y)$ is any function of x and y. If we continue sampling the random variables, finding a sequence of pairs of values (x_i, y_i), $i = 1, \ldots, N$, we can calculate the average

$$f_{\text{av}} = \frac{1}{N} \sum_{i=1}^{N} f(x_i, y_i). \quad (1.36)$$

The limiting value of this average of $f(x, y)$ as the number of pairs N goes to infinity is the expectation value $\langle f(x, y) \rangle$. The formula for the expectation value of $f(x, y)$ can be surmised:

$$\langle f(x, y) \rangle = \int_{-\infty}^{\infty} dx \int_{-\infty}^{\infty} dy\, \rho(x, y)\, f(x, y). \quad (1.37)$$

Expectation is an alternative name for *mean* and both will be used interchangeably in this block.

This expression may be derived by dividing the (x, y)-plane into small rectangular elements, each of size $\delta x \times \delta y$. The probability of x and y being within the element centred at coordinates (x,y) is $\delta P \simeq \rho(x,y)\,\delta x\,\delta y$. The mean value of $f(x,y)$ is thus

$$\langle f(x,y) \rangle = \sum_{(x,y)} f(x,y)\,\delta P \simeq \sum_{(x,y)} \rho(x,y)\,f(x,y)\,\delta x\,\delta y, \qquad (1.38)$$

where the sum runs over all the rectangular elements centred at coordinates (x,y). Taking the limit as $\delta x \to 0$, $\delta y \to 0$, the double sum becomes the double integral (1.37).

When dealing with two random variables, a useful concept is the *correlation*, which measures the extent to which they tend to be related. For example, the heights and weights of people are two different random variables, but it is expected that tall people are more likely to be of above average weight. These variables would be described as 'correlated'. The tendency of two variables to be correlated is described by means of a *correlation coefficient*. The correlation coefficient of two random variables x and y is defined as the mean value of the product of the deviation of each random variable from its mean value. Expressed in terms of symbols, the correlation coefficient is given by

$$C_{xy} = \langle \Delta x \Delta y \rangle, \quad \text{where} \quad \Delta x = x - \langle x \rangle,\ \Delta y = y - \langle y \rangle. \qquad (1.39)$$

Note that the quantity C_{xy} is positive if large values of x tend to occur at the same time as large values of y, and negative if large values of x tend to occur along with small values of y.

The correlation coefficient may be expressed in a form which is more convenient for calculation, as follows. We write the expectation value in terms of an integral using equation (1.37), and simplify the result using equation (1.35) and a similar expression for $\langle y \rangle$:

$$\begin{aligned}
C_{xy} &= \int_{-\infty}^{\infty} dx \int_{-\infty}^{\infty} dy\,[xy - \langle y \rangle x - \langle x \rangle y + \langle x \rangle \langle y \rangle]\,\rho(x,y) \\
&= \langle xy \rangle - \langle y \rangle \int_{-\infty}^{\infty} dx \int_{-\infty}^{\infty} dy\,x\,\rho(x,y) \\
&\quad - \langle x \rangle \int_{-\infty}^{\infty} dx \int_{-\infty}^{\infty} dy\,y\,\rho(x,y) + \langle x \rangle \langle y \rangle \\
&= \langle xy \rangle - \langle x \rangle \langle y \rangle. \qquad (1.40)
\end{aligned}$$

Exercise 1.21

Show that $C_{xy} = 0$ if x and y are independent.

Exercise 1.22

Verify that $C_{xx} = \text{Var}(x)$.

The correlation coefficient is defined in the same way for discrete random variables. Consider the case where the random variables x and y take discrete values X_i, $i = 1, \ldots, n$, and Y_j, $j = 1, \ldots, m$, with the joint probability of observing X_i and Y_j being $P_{i,j}$.

In this case, the expression for $\langle xy \rangle$ is

$$\langle xy \rangle = \sum_{i=1}^{n} \sum_{j=1}^{m} P_{i,j} X_i Y_j$$

1.2 Statistics

and the correlation coefficient is given by

$$\begin{aligned}C_{xy} &= \langle xy \rangle - \langle x \rangle \langle y \rangle \\ &= \sum_{i=1}^{n}\sum_{j=1}^{m} P_{i,j} X_i Y_j - \left[\sum_{i=1}^{n}\sum_{j=1}^{m} P_{i,j} X_i\right] \times \left[\sum_{i=1}^{n}\sum_{j=1}^{m} P_{i,j} Y_j\right].\end{aligned} \quad (1.41)$$

Exercise 1.23

Determine $\langle xy \rangle$ where x and y are the numbers uppermost on two successive throws of a die.

One further general result, which will be required in Chapter 2, concerns the sum X of M random variables x_i, $i = 1, \ldots, M$:

$$X = \sum_{i=1}^{M} x_i. \quad (1.42)$$

In general, these random variables have a joint probability density function $\rho(x_1, x_2, \ldots, x_M)$, such that the element of probability δP for finding the first variable between x_1 and $x_1 + \delta x_1$, the second between x_2 and $x_2 + \delta x_2$, etc., is

$$\delta P = \rho(x_1, x_2, \ldots, x_M)\, \delta x_1\, \delta x_2 \cdots \delta x_M. \quad (1.43)$$

The mean value of the sum of these random variables is given by a simple relation, namely the sum of the mean values of each variable:

$$\langle X \rangle = \left\langle \sum_{i=1}^{M} x_i \right\rangle = \sum_{i=1}^{M} \langle x_i \rangle. \quad (1.44)$$

This is a very useful relationship. To understand how it is obtained, note that the mean value of each variable is given by a generalisation of equation (1.35):

$$\langle x_i \rangle = \int_{-\infty}^{\infty} dx_1 \int_{-\infty}^{\infty} dx_2 \cdots \int_{-\infty}^{\infty} dx_M\, x_i\, \rho(x_1, x_2, \ldots, x_M). \quad (1.45)$$

Equation (1.44) can then be obtained as follows:

$$\begin{aligned}\langle X \rangle &= \int_{-\infty}^{\infty} dx_1 \int_{-\infty}^{\infty} dx_2 \cdots \int_{-\infty}^{\infty} dx_M\, (x_1 + x_2 + \cdots + x_M) \\ &\qquad\qquad \times \rho(x_1, x_2, \ldots, x_M) \\ &= \sum_{i=1}^{M} \int_{-\infty}^{\infty} dx_1 \int_{-\infty}^{\infty} dx_2 \cdots \int_{-\infty}^{\infty} dx_M\, x_i\, \rho(x_1, x_2, \ldots, x_M) \\ &= \sum_{i=1}^{M} \langle x_i \rangle.\end{aligned} \quad (1.46)$$

In the case where the variables x_i are independent, the joint probability density factorises as $\rho(x_1, x_2, \ldots, x_M) = \rho_1(x_1)\, \rho_2(x_2) \cdots \rho_M(x_M)$, and the integrals in equation (1.46) factorise to give

$$\begin{aligned}\langle x_i \rangle &= \int_{-\infty}^{\infty} dx_1\, \rho_1(x_1) \int_{-\infty}^{\infty} dx_2\, \rho_2(x_2) \cdots \int_{-\infty}^{\infty} dx_i\, x_i\, \rho_i(x_i) \\ &\qquad \cdots \int_{-\infty}^{\infty} dx_M\, \rho_M(x_M) \\ &= \int_{-\infty}^{\infty} dx_i\, x_i\, \rho_i(x_i).\end{aligned} \quad (1.47)$$

The final equality uses the fact that all of the integrals except one are normalised, according to equation (1.14). Thus the mean of each variable is determined by a one-dimensional integral, which simplifies the calculation considerably. Actually, this property holds quite generally, for non-independent as well as independent variables, but the reason why it also holds for non-independent variables is more complicated and will not be explored further. In any case, this more general situation will not arise elsewhere in this block; only independent variables will be encountered.

The evaluation of the mean value of the sum by performing multiple integrals is impractical in most cases. However, we have seen that this quantity can be calculated very easily if the mean values of each component are known. This illustrates a general principle, that when dealing with multiple random variables it is usually more efficient to work in terms of statistics rather than probability densities.

1.3 The normal distribution

In Block 0 we considered functions of the form

$$f(x) = A \exp[-\alpha(x - x_0)^2], \tag{1.48}$$

where A, x_0 and α are constants with $\alpha > 0$, which are known as *Gaussian* functions. A random variable having a probability density which is a Gaussian function (with the coefficients A, x_0 and α real) is said to have a *Gaussian distribution*, or a *normal distribution*.

The terms 'Gaussian' and 'normal' in this context are used interchangeably.

Many quantities are found to have probability densities which are well approximated by the normal distribution. Figure 1.4 is a plot of the probability density for weights (in kilograms) of packages containing six apples, displayed as a histogram. It is compared with a smooth curve, which is a good match to the data. The probability density plotted as the smooth curve is a normal distribution, and it clearly provides a very good description of the data.

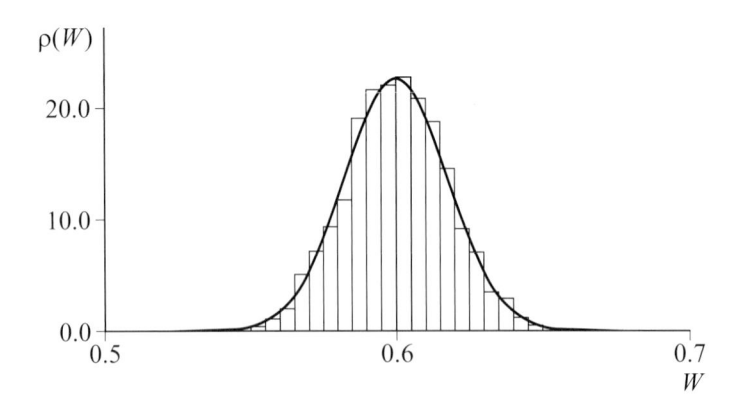

Figure 1.4 A histogram showing the distribution of weights of 2000 packages each containing six apples. The continuous curve is a normal distribution, for comparison. (For the simulated data used to create this example, the parameters occurring in equation (1.48) are $\alpha = 1600$, $x_0 = 0.6$, $A = 40/\sqrt{\pi}$.)

1.3 The normal distribution

You may not be familiar with histograms, so let us pause to discuss the construction of Figure 1.4. The range of possible weights is divided into small intervals (in this case, 40 intervals of width 0.005), and the number of weights in each interval is recorded. The plot consists of rectangular bars. The width of each bar indicates the extent of each corresponding interval, and the height of the bar is proportional to the number of weights N_i recorded in that interval. This is the standard way of presenting empirical information about a probability distribution. The vertical scale has been chosen so that the total area of all of the rectangles is unity, so that the histogram approximates the normalised probability density. The heights of the columns therefore approximate the probability density function, $\rho(x)$, which satisfies equation (1.14).

As the name suggests, normal distributions occur very frequently. The correspondence clearly cannot be exact in our example because the weight of a bag of apples cannot be negative, whereas the function (1.48) remains positive when x is negative. Note, however, that in this case, as can be seen from Figure 1.4, the probability density function modelled by the Gaussian function, $f(W)$, takes extremely small values when $W < 0$. In fact, the normal distribution is often a very good approximation.

The reason for the ubiquity of normal distributions will be discussed in Chapter 5, on the central limit theorem.

Calculations involving normal distributions often require their moments, which can be obtained from integrals of the form

$$I_n = \int_{-\infty}^{\infty} dx\, x^n \exp(-x^2). \tag{1.49}$$

(We shall confine our attention to the case when $n \geq 0$ is an even integer; when n is odd, the integral is zero because the integrand is an odd function.) The first of these integrals was discussed in Block 0:

$$I_0 = \int_{-\infty}^{\infty} dx\, \exp(-x^2) = \sqrt{\pi}. \tag{1.50}$$

The value of I_2 is obtained in the following exercise.

Exercise 1.24

Show that $I_2 = \sqrt{\pi}/2$. [Hint: Use integration by parts.] You may assume that $x^n \exp(-x^2) \to 0$ as $|x| \to \infty$.

Using integration by parts, it can be shown that the integrals I_n satisfy the recursion relation

$$I_n = \frac{n-1}{2} I_{n-2}. \tag{1.51}$$

Exercise 1.25

Verify relation (1.51).

Exercise 1.26

Show that if the Gaussian probability density function (1.48) is normalised (satisfying equation (1.14) with f replacing ρ), A and α are related by $A = \sqrt{\alpha/\pi}$.

Exercise 1.27

Show that the mean and variance of the normal distribution are $\langle x \rangle = x_0$ and $\sigma^2 = 1/2\alpha$, respectively.

Using the integrals obtained above, it can be verified that if σ is the standard deviation of a normal distribution with mean value x_0, the probability density function may be written in the form

$$\rho(x) = \frac{1}{\sqrt{2\pi}\sigma} \exp[-(x - x_0)^2/2\sigma^2]. \tag{1.52}$$

This is a very useful standard form for the normal probability density function.

Example 1.4

Verify that equation (1.52) does indeed correctly describe the normal distribution with mean x_0 and standard deviation σ.

Solution

Equation (1.52) is of the same form as equation (1.48), with $\alpha = 1/2\sigma^2$ and $A = 1/\sqrt{2\pi}\sigma$. Using the result of Exercise 1.26, the distribution is normalised because the factors A and α are related by $A = \sqrt{\alpha/\pi}$. Also, the results of Exercise 1.27 show that the mean value is $\langle x \rangle = x_0$, and the variance is $1/2\alpha = \sigma^2$. The density (1.52) does therefore represent a normalised distribution with mean x_0 and standard deviation σ. ∎

Exercise 1.28

If x has a Gaussian probability density, show that $\langle (x - x_0)^4 \rangle = 3\sigma^4$.

1.3.1 The error function

We have already seen that probabilities are determined by integrals of probability density functions using expressions such as (1.12). The integral of the Gaussian function is therefore a very significant object in probability theory. This integral cannot be expressed exactly in terms of elementary functions, and is accordingly one of a set of functions known as the *special functions*. The *error function* is a definite integral of a Gaussian function, written $\text{erf}(x)$, defined by

$$\text{erf}(x) = \frac{2}{\sqrt{\pi}} \int_0^x dy \exp(-y^2). \tag{1.53}$$

The interpretation of the error function is as follows. The probability of a normally distributed variable differing from the mean by less than $\lambda\sigma$ is $\text{erf}(\lambda/\sqrt{2})$. This is demonstrated by the following calculation:

$$P(x_0 - \lambda\sigma, x_0 + \lambda\sigma) = \frac{1}{\sqrt{2\pi}\sigma} \int_{x_0 - \lambda\sigma}^{x_0 + \lambda\sigma} dx \exp[-(x - x_0)^2/2\sigma^2]$$

$$= \frac{1}{\sqrt{2\pi}\sigma} \int_{-\lambda\sigma}^{\lambda\sigma} du \exp[-u^2/2\sigma^2]$$

$$= \frac{1}{\sqrt{\pi}} \int_{-\lambda/\sqrt{2}}^{\lambda/\sqrt{2}} dy \exp(-y^2) = \text{erf}(\lambda/\sqrt{2}), \tag{1.54}$$

1.3 The normal distribution

where we used the changes of variables $u = x - x_0$ and $y = u/\sqrt{2}\sigma$, and the final step uses the fact that $\exp(-y^2)$ is an even function. The error function is plotted in Figure 1.5 for $x \geq 0$.

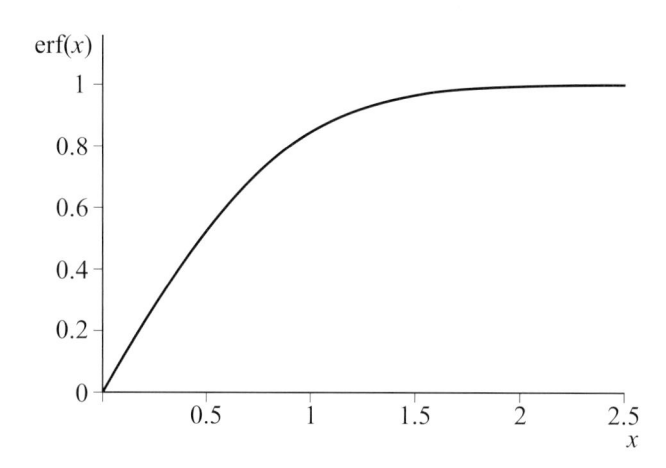

Figure 1.5 The error function $\text{erf}(x)$ for $x \geq 0$

You may also see expressions containing the *complementary error function*, $\text{erfc}(x)$, defined by $\text{erfc}(x) = 1 - \text{erf}(x)$.

Another notation for an integral of a Gaussian function is

$$N(\lambda) = \frac{1}{\sqrt{2\pi}} \int_{-\infty}^{\lambda} dy \exp(-y^2/2), \qquad (1.55)$$

which is sometimes called the normal distribution function. The integral $N(\lambda)$ is the probability that a Gaussian random variable is less than $x_0 + \lambda\sigma$ (where x_0 and σ are, respectively, the mean and standard deviation). Note that $N(\infty) = 1$, as required since the probability that a Gaussian random variable is less than ∞ is obviously one. Figure 1.6 is a graph of $N(\lambda)$. Comparison with equation (1.54) shows that $N(\lambda) = \frac{1}{2}\text{erf}(\lambda/\sqrt{2}) + \frac{1}{2}$. This can be seen by first noting that the substitution $y = u/\sqrt{2}$ in the final integral in the derivation (1.54) implies that $\text{erf}(\lambda/\sqrt{2}) = N(\lambda) - N(-\lambda)$. Secondly, as will be seen from the hint in Exercise 1.29 below, we have $N(-\lambda) = 1 - N(\lambda)$. Combining these two results gives the required expression.

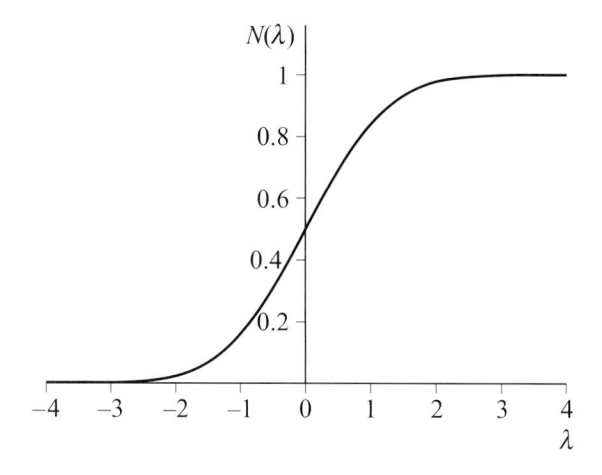

Figure 1.6 The function $N(\lambda)$ is the probability that a normally distributed variable is less than $x_0 + \lambda\sigma$ (where x_0 and σ are the mean and standard deviation)

Both the function $N(\lambda)$ and the error function will be used in later chapters. Values of $N(\lambda)$ are tabulated below.

Table 1.2

λ	$N(\lambda)$
-4	3.167×10^{-5}
-3	$0.001\,350$
-2.5	$0.006\,210$
-2	$0.022\,76$
-1.5	$0.066\,81$
-1	$0.158\,7$
-0.5	$0.308\,5$
0	$0.500\,0$

Exercise 1.29

Calculate $N(1)$ and $N(2)$ using the data in the table above. [Hint: Note that

$$N(-x) = \frac{1}{\sqrt{2\pi}} \int_{-\infty}^{-x} dy \exp(-y^2/2)$$
$$= \frac{1}{\sqrt{2\pi}} \int_{x}^{\infty} du \exp(-u^2/2) = 1 - N(x),$$

where the change of variable $y = -u$ was used in the second integral, and the fact that $N(\infty) = 1$ was used in the last equality.]

Exercise 1.30

What is the probability that a Gaussian random variable does not exceed its mean value by 2.5 multiples of its standard deviation? What is the probability that a Gaussian random variable does not differ from its mean value by more than 3 multiples of its standard deviation?

1.4 Outcomes

After studying this chapter you should be able to:
- understand the concept of probability for discrete events;
- determine probabilities for combinations of mutually exclusive events;
- determine probabilities for successive independent trials;
- understand the notion of a continuous random variable and its description using the probability density function;
- understand the joint probability of many discrete or continuous random variables;
- understand the joint probability density function for two or more continuous random variables;
- calculate the average of observations of a single random variable, and understand its relationship to the mean or expectation value of that

random variable;
- calculate moments, the variance and the standard deviation of a random variable;
- calculate the correlation coefficient of two random variables;
- understand the normal (or Gaussian) distribution and its probability density function, as well as the error function and normal distribution function;
- calculate moments of Gaussian random variables.

1.5 Further Exercises

These exercises are optional.

The following exercises are harder than those in the main text. The results are not used in subsequent chapters, but they will broaden and deepen your knowledge of probability.

Exercise 1.31

If every couple were to continue having children until they have one boy, what would be the ratio of males to females in the next generation? What would be the mean number of children per couple?

Assume that everyone enters into one relationship that produces offspring, and that all couples are fertile and can produce an arbitrary number of offspring. Also, ignore the existence of twins, triplets, etc.; i.e. assume that all pregnancies produce just one child at a time.

[Hints: Using the obvious notation B = boy, G = girl, the probability of producing one child (B) is $\frac{1}{2}$, and the probability of producing two children (G, B) is $\frac{1}{4}$. The probability of producing $n+1$ children (nG, B) is $2^{-(n+1)}$. The mean number of girls per couple is

$$\langle n \rangle = \sum_{n=1}^{\infty} n \left(\tfrac{1}{2}\right)^{n+1}.$$

To find this sum, introduce the geometric series (see Block 0, equation (1.2))

$$S(x) = \sum_{n=1}^{\infty} x^{-n} = \frac{x^{-1}}{1 - x^{-1}} = \frac{1}{x - 1},$$

and note that the derivative of this sum is in the form of the series required:

$$-\frac{d}{dx} S(x) = \sum_{n=1}^{\infty} \frac{n}{x^{n+1}}.$$

Setting $x = 2$, you should conclude that the gender balance and the size of the population are maintained.]

Exercise 1.32

Often we might be interested in a series of random events which occur with the same average frequency over any long period of time. Examples are the occurrences of earthquakes, and the series of clicks from a Geiger counter observing radioactive decay. We would like to know the distribution of times between these events. In the case where the events are independent of each other, this distribution can be calculated by the following approach.

The probability that an event occurs in a very short interval of time, δt, is assumed to be $R\,\delta t$, where R is known as the *rate constant*. Let $P(t)$ be the probability that *no* event will have occurred after time t. Because the events are independent, the probability that no event will have occurred after time $t + \delta t$ is $P(t)$ multiplied by the probability of a small time interval δt passing with no event, namely $1 - R\,\delta t$. It follows that

$$P(t + \delta t) = P(t)(1 - R\,\delta t).$$

(a) Deduce a differential equation for $P(t)$, and show that

$$P(t) = \exp(-Rt).$$

This is known as the *Poisson distribution*.

Show that the mean time between events is $\langle t \rangle = 1/R$.

(b) A cyclist travels 20 km per day, and notices that she suffers 15 punctures in 100 days. Estimate the probability that she will complete a 600 km cycling holiday without a puncture.

Solutions to Exercises in Chapter 1

Solution 1.1

No solution is given, as your results will differ in detail from anyone else's.

Solution 1.2

The number of picture cards is 12 (a King, a Queen and a Jack in each of the 4 suits). The probability of drawing any one of 52 cards is 1/52. The probability of drawing any picture card is $12 \times \frac{1}{52} = 3/13$.

Solution 1.3

The total number of deaths was 575 200. The table enables us to estimate the probability of death at ages 25–44 as $P_{25-44} = 14\,600/575\,200 \simeq 0.0254$. Similarly, $P_{45-64} = 88\,200/575\,200 \simeq 0.1533$. The events corresponding to death occurring in either of these age ranges are mutually exclusive, so the probability of death occurring in the age range 25–64 is the sum of these values:

$$P_{25-64} \simeq 0.0254 + 0.1533 \simeq 0.179.$$

(Alternatively, the total number of deaths at ages between 25 and 64 is $N_{25-64} = N_{25-44} + N_{45-64} = 14\,600 + 88\,200 = 102\,800$. Hence $P_{25-64} = 102\,800/575\,200 \simeq 0.179$.)

This is a good estimate of the probability that a death occurring in the next year will occur between ages 25 and 64. Factors influencing longevity (such as medical advances and wars) change over time, so this may not be a good estimate that someone born this year will die between ages 25 and 64.

Solution 1.4

On throwing a single die, the probability of obtaining a 1 or a 2 is $P_{1 \text{ or } 2} = \frac{2}{6} = \frac{1}{3}$. The probability of throwing two dice with each showing less than three is $P = \frac{1}{3} \times \frac{1}{3} = \frac{1}{9}$.

As a check, consider the possible combinations of two throws. There are $6 \times 6 = 36$ possible combinations. Of these, only the following four have no number greater than two: $(1,1), (1,2), (2,1), (2,2)$. The required probability is $\frac{4}{36} = \frac{1}{9}$, confirming the result obtained from equation (1.8).

Solution 1.5

The weather is unrelated to the convention determining which days of the week are available for leisure, so the formula for independent events (equation (1.8)) is applicable. The probability of a day falling on a weekend is 2/7. Thus the probability of a day falling on a weekend and also being suitable for the sport is $4/10 \times 2/7 = 8/70 \simeq 0.114$.

The number of days available for this sport every year is therefore approximately 365×0.114, that is, approximately 42 days per year.

Solution 1.6

A total of 7 can be obtained in 6 ways: $1+6, 2+5, 3+4, 4+3, 5+2, 6+1$. There are $6 \times 6 = 36$ possible outcomes from throwing the two dice. Since each possible outcome has the same probability, each combination of numbers has probability 1/36. There are six different outcomes for which the total is 7, and these are mutually exclusive, so the probabilities add. The probability of the total being equal to 7 is therefore $6/36 = 1/6$.

Solution 1.7

Let the constant in equation (1.16) be denoted by C. The normalisation condition (1.14) is

$$\int_{-\infty}^{\infty} dx\, \rho(x) = C \int_{x_1}^{x_2} dx\, 1 = C(x_2 - x_1) = 1,$$

so $C = 1/(x_2 - x_1)$.

Solution 1.8

The angle at which the tyre valve comes to rest when the aircraft is parked is random. Since the wheel is circular, no direction is preferred, and the angle θ with respect to the vertical should have a uniform distribution $\rho(\theta) = \text{constant} = 1/360$, in the range $-180° < \theta < 180°$.

The probability that the angle is between $\theta_1 = -20°$ and $\theta_2 = +20°$ is

$$P = \int_{\theta_1}^{\theta_2} d\theta\, \rho(\theta) = \frac{1}{360} \int_{-20}^{+20} d\theta = \frac{40}{360} = \frac{1}{9}.$$

The probability that the aircraft will have to be moved to check the tyre pressure is therefore $\frac{1}{9}$.

Solution 1.9

(a) Since $\rho(t) = 0$ for $t < 0$, the normalisation condition (1.14) is

$$1 = \int_0^{\infty} dt\, \rho(t) = A \int_0^{\infty} dt\, \exp(-t/t_0) = At_0.$$

The substitution $u = t/t_0$ was used to evaluate the integral.

The probability density is therefore correctly normalised by setting $A = 1/t_0$.

(b) The probability P of waiting longer than a time T is the area under the graph of $\rho(t)$ between T and ∞:

$$P = \frac{1}{t_0} \int_T^{\infty} dt\, \exp(-t/t_0) = \exp(-T/t_0).$$

The probability of waiting longer than $T = 45$ minutes is therefore

$$P = \exp(-45/15) = \exp(-3) \simeq 0.049.$$

There is roughly a one in twenty chance that the passenger will walk home if the buses are running normally.

Solution 1.10

The joint probability density is

$$\rho(t_1, t_2) = \frac{1}{t_0} \exp(-t_1/t_0) \times \frac{1}{t_0} \exp(-t_2/t_0) = \frac{\exp[-(t_1 + t_2)/t_0]}{t_0^2},$$

where $t_0 = 15$ minutes. The probability P that both waiting times are less than t_0 is obtained by integrating over the square domain $0 < t_1 < t_0$, $0 < t_2 < t_0$:

$$P = \int_0^{t_0} dt_1 \int_0^{t_0} dt_2\, \rho(t_1, t_2)$$

$$= \frac{1}{t_0^2} \int_0^{t_0} dt_1 \exp(-t_1/t_0) \int_0^{t_0} dt_2 \exp(-t_2/t_0)$$

$$= \frac{1}{t_0} \int_0^{t_0} dt_1 \exp(-t_1/t_0)[1 - \exp(-1)]$$

$$= [1 - \exp(-1)]^2.$$

The substitution $u = t/t_0$ was used to simplify each integral.

Solution 1.11

The joint probability density is the same as in Exercise 1.10, and in this case the region of integration is a triangle in the (t_1, t_2)-plane where $t_1 > 0$, $t_2 > 0$ and $t_1 + t_2 < t_0$ (see Figure 1.7).

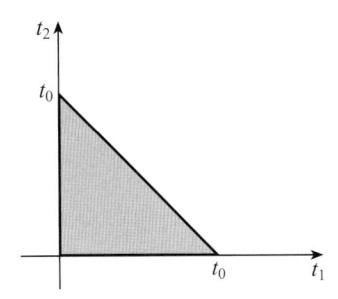

Figure 1.7 The region of integration considered in Exercise 1.11

The probability P that $t_1 + t_2 < t_0$ is given by

$$P = \frac{1}{t_0^2} \int_0^{t_0} dt_1 \exp(-t_1/t_0) \int_0^{t_0-t_1} dt_2 \exp(-t_2/t_0)$$
$$= \frac{1}{t_0} \int_0^{t_0} dt_1 \exp(-t_1/t_0) \left[-\exp(-u)\right]_0^{1-t_1/t_0}$$
$$= \int_0^1 dv \exp(-v)[1 - \exp(-1+v)]$$
$$= 1 - 2\exp(-1).$$

Using substitutions
$u = t_2/t_0$
$v = t_1/t_0$.

Solution 1.12

The probability of the random variables x and y taking any values in the (x, y)-plane is unity. Setting the probability in equation (1.19) equal to unity when $\mathcal{A}$ is the entire (x, y)-plane gives the normalisation condition

$$\int_{-\infty}^{\infty} dx \int_{-\infty}^{\infty} dy\, \rho(x, y) = 1.$$

Also, equation (1.18) implies that the joint probability density can never be negative.

Solution 1.13

Let δP be the probability that the first variable is between x and $x + \delta x$. Also, let j be an integer, and let δP_j be the probability that both variables are in the respective small intervals $[x, x+\delta x]$, $[j\,\delta y, (j+1)\,\delta y]$. Note that $\delta P \simeq \rho_1(x)\,\delta x$ and $\delta P_j \simeq \rho(x, j\,\delta y)\,\delta x\,\delta y$, and that $\delta P = \sum_j \delta P_j$, so

$$\rho_1(x)\,\delta x \simeq \sum_{j=-\infty}^{\infty} \rho(x, j\,\delta y)\,\delta x\,\delta y.$$

In the limit as $\delta y \to 0$, the sum becomes an integral, so the probability density for x alone is the integral of the joint probability density over the other variable:

$$\rho_1(x) = \int_{-\infty}^{\infty} dy\, \rho(x, y).$$

When x and y are independent variables, we have $\rho(x, y) = \rho_1(x)\rho_2(y)$, where $\rho_1(x)$ and $\rho_2(y)$ are the probability densities of x and y, respectively. In this case, the integral gives

$$\rho_1(x) = \int_{-\infty}^{\infty} dy\, \rho_1(x)\,\rho_2(y)$$
$$= \rho_1(x) \int_{-\infty}^{\infty} dy\, \rho_2(y) = \rho_1(x),$$

The final step used the fact that $\rho_2(y)$ is normalised, satisfying equation (1.14).

as expected.

Solution 1.14

For throwing a single die, there are 6 possible outcomes, where the number on top takes the values $j = 1$ to $j = 6$, all with probability $P_j = 1/6$. The mean value of $x = j$ is

$$\langle x \rangle = \sum_{j=1}^{6} P_j\, j = \frac{1}{6} \sum_{j=1}^{6} j = \frac{1+2+3+4+5+6}{6} = \frac{21}{6} = 3.5.$$

Note that the average of the second sequence in Example 1.3 is much closer to this mean value than is the average of the first (shorter) sequence.

Solution 1.15

Assuming that $x_2 > x_1$, the probability density is $1/(x_2 - x_1)$ in the interval $[x_1, x_2]$, and zero outside it (see Exercise 1.7). Then integral (1.30) becomes

$$\langle x \rangle = \frac{1}{x_2 - x_1} \int_{x_1}^{x_2} dx\, x$$

$$= \frac{1}{2(x_2 - x_1)} \left[x^2 \right]_{x_1}^{x_2}$$

$$= \frac{x_2^2 - x_1^2}{2(x_2 - x_1)} = \tfrac{1}{2}(x_2 + x_1).$$

Solution 1.16

Remembering that the density $\rho(x)$ is zero for $x < 0$, the mean value is

$$\langle x \rangle = \frac{1}{x_0} \int_0^\infty dx\, x \exp(-x/x_0)$$

$$= x_0 \int_0^\infty du\, u \exp(-u)$$

$$= x_0 \left[-u \exp(-u) \right]_0^\infty + x_0 \int_0^\infty du \exp(-u)$$

$$= (x_0 \times 0) + (x_0 \times 1),$$

so $\langle x \rangle = x_0$.

The calculation used the change of variable $u = x/x_0$, followed by an integration by parts.

Solution 1.17

The probability density function is

$$\rho(x) = \begin{cases} 1, & x \in [0,1], \\ 0, & \text{otherwise.} \end{cases}$$

The moments are therefore

$$M_k = \int_{-\infty}^{\infty} dx\, x^k \rho(x) = \int_0^1 dx\, x^k = \frac{1}{k+1}.$$

Solution 1.18

In the case of a discrete distribution, by analogy with equation (1.28) the expression for $\langle x^k \rangle$ is

$$\langle x^k \rangle = \sum_{j=1}^{m} X_j^k P_j.$$

If x is the number obtained from a throw of a die, this takes discrete values from $j = 1$ to $j = 6$, with $P_j = \tfrac{1}{6}$. So the mean value of x^2 is

$$\langle x^2 \rangle = \sum_{j=1}^{6} \tfrac{1}{6} j^2 = \frac{1+4+9+16+25+36}{6} = \frac{91}{6} = 15.1\dot{6}.$$

Solution 1.19

From the result of Exercise 1.17, the first and second moments are $M_1 = \frac{1}{2}$, $M_2 = \frac{1}{3}$. The variance is therefore $\mathrm{Var}(x) = M_2 - M_1^2 = \frac{1}{12}$.

Solution 1.20

The expression for the mean value of x in terms of its probability density $\rho_1(x)$ is

$$\langle x \rangle = \int_{-\infty}^{\infty} dx\, x\, \rho_1(x),$$

using the notation from Exercise 1.13. The probability density $\rho_1(x)$ is given in terms of $\rho(x,y)$ in the solution to Exercise 1.13. Inserting this expression gives

$$\langle x \rangle = \int_{-\infty}^{\infty} dx \int_{-\infty}^{\infty} dy\, x\, \rho(x,y).$$

Solution 1.21

If x and y are independent, their joint probability density factorises, and may be written as $\rho(x,y) = \rho_1(x)\rho_2(y)$ where ρ_1 and ρ_2 are the probability densities for x and y, respectively. The mean value of the product xy is then

$$\langle xy \rangle = \int_{-\infty}^{\infty} dx \int_{-\infty}^{\infty} dy\, xy\, \rho(x,y)$$
$$= \int_{-\infty}^{\infty} dx\, x\, \rho_1(x) \int_{-\infty}^{\infty} dy\, y\, \rho_2(y)$$
$$= \langle x \rangle \langle y \rangle.$$

So the correlation coefficient is $C_{xy} = \langle xy \rangle - \langle x \rangle \langle y \rangle = 0$.

Solution 1.22

Comparing equations (1.39) and (1.32), we have

$$C_{xx} = \langle (x - \langle x \rangle)(x - \langle x \rangle) \rangle = \mathrm{Var}(x).$$

Alternatively, comparing equations (1.40) and (1.34), we have $C_{xx} = M_2 - M_1^2 = \mathrm{Var}(x)$.

Solution 1.23

For two successive throws, yielding values i on the first throw and j on the second, all combinations of two numbers have equal probability, namely $P_{i,j} = 1/36$. The mean value of the product $xy = i \times j$ is

$$\langle xy \rangle = \sum_{i,j} P_{i,j}\, i \times j = \tfrac{1}{36} \sum_{i=1}^{6} \sum_{j=1}^{6} i \times j$$
$$= \tfrac{1}{36} \sum_{i=1}^{6} i \sum_{j=1}^{6} j$$
$$= \tfrac{1}{36} \times (21)^2 = 12.25.$$

Solution 1.24

Using the substitutions $u = x$ and $dv/dx = x\exp(-x^2)$ (so $v = -\exp(-x^2)/2$) in the formula for integration by parts,

$$\int dx\, u \frac{dv}{dx} = [uv] - \int dx\, v \frac{du}{dx},$$

gives

$$I_2 = \int_{-\infty}^{\infty} dx\, x^2 \exp(-x^2) = \left[\frac{-x\exp(-x^2)}{2}\right]_{-\infty}^{\infty} + \int_{-\infty}^{\infty} dx\, \frac{\exp(-x^2)}{2}$$
$$= 0 + \tfrac{1}{2} I_0 = \frac{\sqrt{\pi}}{2}.$$

Solution 1.25

Using the substitutions $u = x^{n-1}$ and $dv/dx = x\exp(-x^2)$ in the formula for integration by parts gives

$$I_n = \int_{-\infty}^{\infty} dx\, x^n \exp(-x^2)$$
$$= \left[\frac{-x^{n-1}\exp(-x^2)}{2}\right]_{-\infty}^{\infty} + \frac{n-1}{2}\int_{-\infty}^{\infty} dx\, x^{n-2}\exp(-x^2)$$
$$= \frac{n-1}{2} I_{n-2}.$$

Solution 1.26

If the probability density is normalised, we have

$$\int_{-\infty}^{\infty} dx\, A \exp[-\alpha(x-x_0)^2] = 1.$$

Using the change of variable $u = \sqrt{\alpha}(x-x_0)$, so that $du/dx = \sqrt{\alpha}$, we obtain

$$A\int_{-\infty}^{\infty} dx \exp[-\alpha(x-x_0)^2] = \frac{A}{\sqrt{\alpha}}\int_{-\infty}^{\infty} du \exp(-u^2) = A\sqrt{\frac{\pi}{\alpha}}.$$

Thus we must have $A = \sqrt{\alpha/\pi}$ in order for the probability density function to be normalised.

Solution 1.27

To calculate the mean value, use the substitution $u = \sqrt{\alpha}(x-x_0)$:

$$\langle x \rangle = \int_{-\infty}^{\infty} dx\, x\, \rho(x)$$
$$= \sqrt{\frac{\alpha}{\pi}} \int_{-\infty}^{\infty} dx\, x \exp[-\alpha(x-x_0)^2]$$
$$= \sqrt{\frac{\alpha}{\pi}} \int_{-\infty}^{\infty} \frac{du}{\sqrt{\alpha}} \left(\frac{u}{\sqrt{\alpha}} + x_0\right) \exp(-u^2)$$
$$= \frac{1}{\sqrt{\pi\alpha}} \int_{-\infty}^{\infty} du\, u \exp(-u^2) + \frac{x_0}{\sqrt{\pi}} \int_{-\infty}^{\infty} du \exp(-u^2)$$
$$= \frac{0}{\sqrt{\pi\alpha}} + \frac{x_0\sqrt{\pi}}{\sqrt{\pi}}$$
$$= x_0.$$

The first integral in the fourth line vanishes because the integrand is odd.

The same change of variable helps to calculate the variance:

$$\sigma^2 = \langle (x - \langle x \rangle)^2 \rangle$$
$$= \langle (x - x_0)^2 \rangle$$
$$= \sqrt{\frac{\alpha}{\pi}} \int_{-\infty}^{\infty} dx\, (x-x_0)^2 \exp[-\alpha(x-x_0)^2]$$
$$= \sqrt{\frac{\alpha}{\pi}} \int_{-\infty}^{\infty} \frac{du}{\sqrt{\alpha}} \frac{u^2}{\alpha} \exp(-u^2)$$
$$= \frac{1}{\alpha\sqrt{\pi}} I_2 = \frac{1}{2\alpha},$$

using I_2 as calculated in Exercise 1.24.

Solution 1.28

Using the substitution $u = (x - x_0)/\sqrt{2\sigma^2}$ and relation (1.51), we have

$$\langle (x - x_0)^4 \rangle = \int_{-\infty}^{\infty} dx \, \frac{1}{\sqrt{2\pi}\sigma} (x - x_0)^4 \exp[-(x - x_0)^2/2\sigma^2]$$

$$= \frac{4\sigma^4}{\sqrt{\pi}} \int_{-\infty}^{\infty} du \, u^4 \exp(-u^2)$$

$$= \frac{4\sigma^4}{\sqrt{\pi}} I_4 = \frac{4\sigma^4}{\sqrt{\pi}} \frac{3}{2} I_2 = \frac{6\sigma^4}{\sqrt{\pi}} \frac{\sqrt{\pi}}{2} = 3\sigma^4.$$

Solution 1.29

From the hint we have $N(x) = 1 - N(-x)$. Using the data in the table gives $N(1) = 1 - N(-1) = 1 - 0.1587 = 0.8413$. Similarly, $N(2) = 1 - 0.02276 = 0.97724$.

Solution 1.30

It follows from the description of the function $N(x)$ given beneath equation (1.55) that the probability that a Gaussian random variable does not exceed its mean by more than 2.5 multiples of its standard deviation is $N(2.5) = 1 - N(-2.5)$. From the table, this gives $N(2.5) = 1 - 0.00621 = 0.99379$.

One can see from equation (1.55), and the discussion following it, that the probability that a Gaussian random variable has its value within three standard deviations of its mean is given by $N(3) - N(-3) = 1 - 2N(-3) = 1 - 2 \times 0.00135 = 0.9973$.

Solution 1.31

The probability of producing a boy is $\frac{1}{2}$, and the probability of producing a girl is $\frac{1}{2}$. The probability of producing n girls and 1 boy is $(\frac{1}{2})^n \times \frac{1}{2} = (\frac{1}{2})^{n+1}$. The mean number of girls is

$$\langle n \rangle = \sum_{n=1}^{\infty} n P(n) = \sum_{n=1}^{\infty} n \left(\frac{1}{2}\right)^{n+1}.$$

Now note that this sum is closely related to the derivative of $S(x)$ given in the question:

$$\frac{d}{dx} S(x) = \sum_{n=1}^{\infty} -n \left(\frac{1}{x}\right)^{n+1}$$

$$= \frac{d}{dx} \sum_{n=1}^{\infty} x^{-n}$$

$$= \frac{d}{dx} \left(\frac{1}{x-1}\right) = \frac{-1}{(x-1)^2}.$$

Setting $x = 2$ gives

$$\langle n \rangle = \sum_{n=1}^{\infty} \frac{n}{2^{n+1}} = \frac{1}{(2-1)^2} = 1.$$

The mean number of girls per couple is $\langle n \rangle = 1$, and there is always one boy, so the ratio of male to female births is unity.

Because the mean number of children produced by each couple is two (1 boy and $\langle n \rangle = 1$ girls), the population size is expected to remain constant.

Solution 1.32

(a) Writing
$$P(t+\delta t) = P(t) + P'(t)\,\delta t + O(\delta t^2)$$
and comparing with the expression for $P(t+\delta t)$ given in the question, we see that
$$\frac{dP}{dt} = -RP.$$
Integrating this for $P > 0$ gives
$$\ln P = -Rt + c, \quad \text{or equivalently} \quad P(t) = C\exp(-Rt),$$
where c and C are constants. The constant C is determined by noting that at $t=0$ it is certain that no events have occurred, so $P(0) = 1$, giving $C = 1$.

The mean time between events is
$$\langle t \rangle = \int_0^\infty dt\, t\, \rho(t),$$
where $\rho(t)$ is the probability density for events to be separated by time t. (Note that $\rho(t)$ is zero for negative t, so the lower limit of the integral may be set to zero.) The following argument can be used to determine $\rho(t)$.

First, let $P(t_1, t_2)$ be the probability of observing an event between times t_1 and t_2 (assuming $t_2 \geq t_1 \geq 0$). Note that we have $P(t_1, t_3) = P(t_1, t_2) + P(t_2, t_3)$ (with $t_3 \geq t_2 \geq t_1$). Now, since $P(t) + P(0,t) = 1$, we have already determined that $P(0,t) = 1 - \exp(-Rt)$, so we have
$$P(t_1, t_2) = P(0, t_2) - P(0, t_1)$$
$$= [1 - \exp(-Rt_2)] - [1 - \exp(-Rt_1)]$$
$$= \exp(-Rt_1)\left(1 - \exp[-R(t_2 - t_1)]\right).$$

It follows from the definitions of $\rho(t)$ and $P(t_1, t_2)$ that $P(t, t+\delta t) = \rho(t)\,\delta t + O(\delta t^2)$, so
$$\rho(t) = \lim_{\delta t \to 0} \left[\frac{P(t, t+\delta t)}{\delta t}\right] = R\exp(-Rt).$$

The mean time between events is therefore
$$\langle t \rangle = R\int_0^\infty dt\, t\exp(-Rt) = \frac{1}{R}.$$

(b) The rate at which a puncture occurs is $R = 15/100$ per day. It takes 30 days for the cyclist to travel 600 km given that she cycles 20 km per day. Hence, the probability of *not* getting a puncture after cycling 600 km is
$$P(30) = \exp(-30R) = \exp(-30 \times 15/100) = \exp(-4.5) \simeq 0.011.$$
So the cyclist is very likely to have punctures.

CHAPTER 2
Discrete random functions and random walks

2.1 Introduction

The Introduction to this Block listed some processes and phenomena, illustrated in Figures 0.1 to 0.7, which can be modelled by random functions. This chapter discusses some *discrete* models for these random functions, $f(n)$, which are defined only for integer values of n. For each value of the integer n, the value of the function $f(n)$ is defined by means of a random process (that is, by a sequence of operations which involves random elements, such as tossing coins, or using computer programs which generate apparently random numbers). Later, in Chapter 6, there will be a discussion of how these definitions are extended so that one type of random function, the random walk, can be defined for all real values of its argument.

The concept of a random function is just an extension of the idea of a random number. Random numbers are generated by using a random process to select a number from a specified set (such as the set of real numbers, or the set of positive integers). The random process gives a certain probability for selecting a particular value for the random number (or it gives a probability density, if the numbers are selected from a continuous set). Random functions are constructed in the same manner: a random process has a certain probability (or probability density) for selecting a function from a given set.

In general, it is a difficult task to define sets of functions, and to associate functions in these sets with probabilities. The approach that we shall use in this chapter is to define one very elementary type of random function, which we shall call the *coin-tossing function*. The coin-tossing function is then used to derive two other types of random function, called the *random walk* and the *correlated random function*, which find applications as models of random processes.

The chapter starts by introducing the coin-tossing function in Section 2.2, and this will be used (in Section 2.3) to construct random walks. The random walk is a key concept in this course because of its relation to diffusion processes, and its properties are examined in some detail in Sections 2.4 to 2.7. Finally, Section 2.8 describes the correlated random function.

2.2 An elementary random function

2.2.1 Defining the coin-tossing function

We start by considering the simplest random function, which we call the *coin-tossing function*. It is a mapping from a set of consecutive positive and negative integers $\{-N, \ldots, -2, -1, 0, 1, 2, 3, \ldots, N-1, N\}$ to the set $\{-1, 1\}$, constructed by the following procedure. For each value of n chosen from the first set, toss a coin. If it falls heads up, assign the value $+1$ to the function at n; otherwise, assign $f(n) = -1$. Then pick the next integer n, and repeat for all the elements of the first set. (We could pick the values of n in sequence, but the order in which we take them is, in fact, irrelevant.) By increasing N, this procedure will define the function $f(n)$ for an arbitrarily large range of integers.

The name 'coin-tossing function' is not standard mathematical terminology.

The function will depend upon how the coin falls with each throw, and repeating the entire procedure will produce different examples of the random function. Sometimes these are described as different *realisations*. Two examples (i.e. different realisations) are shown in Figure 2.1. The procedure which generates a realisation of the random function is called a *stochastic process*. We shall use the term 'random function' to refer to both the process that generates the function and one of its realisations. For example, when we refer to the coin-tossing function, it will be clear from the context whether we mean a particular realisation of this random function or the stochastic process which generates such realisations.

The word 'stochastic' has the same meaning as 'random'.

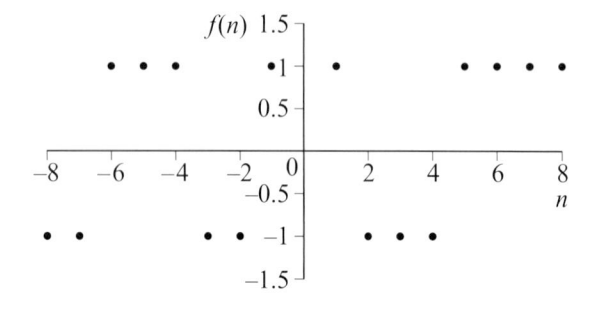

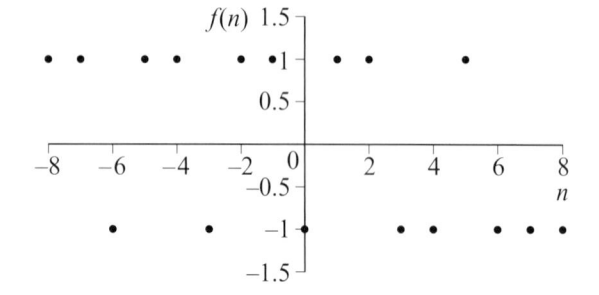

Figure 2.1 Two realisations of the coin-tossing function, defined for the range $n = -8$ to $n = +8$

The coin-tossing function will be used to generate other types of random function. Unless otherwise stated, we shall assume that the limit as $N \to \infty$ is taken, so that $f(n)$ is defined (randomly) for every n.

2.2.2 Statistics of the coin-tossing function

In Chapter 1 we discussed random events in terms of their probabilities, and of course we can discuss random functions using the same approach. For example, given a random function (such as the coin-tossing function) that takes a discrete set of values $f(n)$ on a finite set of points (such as $n = -N, \ldots, 0, 1, \ldots, N$), we can assign a probability for a given realisation of the random function. (In cases where $f(n)$ has a continuous range of values, a probability *density* must be specified.) However, this approach is very cumbersome, because of the large number of variables required. (For example, the coin-tossing function defined for the integers from $-N$ to $+N$ has a probability which is defined in terms of $2N + 1$ random variables, namely the values $f(n)$.) It is preferable to discuss random functions in terms of their statistics. The remainder of this section will describe the statistical properties of the coin-tossing function.

Simple statistics give a clearer understanding of the properties of a random function than can be obtained by considering probabilities. Also, we shall see later that statistics can often be evaluated without writing down probabilities or, for continuous cases, probability density functions. The simplest statistics describing a random function $f(n)$ are the mean $\langle f(n) \rangle$ and the second moment $\langle f^2(n) \rangle$, discussed in Chapter 1, Section 1.2.

We use angular brackets to denote expectation values in the same way as in Chapter 1, but it may be helpful to review what the notation $\langle f(n) \rangle$ means. If we make M observations of a random variable X, then the average of these observations approaches $\langle X \rangle$ in the limit as $M \to \infty$. The quantities $\langle f(n) \rangle$ and $\langle f^2(n) \rangle$ are defined in the same way. If we generate M different realisations of a random function f, then we can determine the average of f for some given choice of the integer argument n, that is, the average of $f(n)$. This average approaches the expectation value $\langle f(n) \rangle$ in the limit as $M \to \infty$. Similarly, if we calculate the average of f^2 at some given value of n over M realisations of the random function, then this average approaches the expectation value $\langle f^2(n) \rangle$ in the limit as $M \to \infty$. There is no fundamental difference between calculating the expectation value of a single random variable, $\langle X \rangle$ say, and the expectation value of a random function with the argument set equal to n, namely $\langle f(n) \rangle$. In both cases, equation (1.28) or (1.30) can be used, depending on whether the quantity that we are averaging has a discrete or a continuous range of values.

In general, both of these expectation values, $\langle f(n) \rangle$ and $\langle f^2(n) \rangle$, will depend upon n. In the case of the coin-tossing function, however, $\langle f(n) \rangle$ and $\langle f^2(n) \rangle$ do not depend upon n, because the procedure used to generate the random number $f(n)$ is the same for all choices of n.

> There is an alternative approach to defining expectation values for random functions. Given a particular realisation of a random function $f(n)$ which, in this case, is defined on all positive integer values n, we could calculate an average over n, such as
>
> $$f_{\text{av}} = \frac{1}{N} \sum_{n=1}^{N} f(n), \qquad (2.1)$$
>
> and consider the limit of this quantity as $N \to \infty$. This average, based on a single realisation, need not be the same as $\langle f(n) \rangle$, which is defined in terms of an average over realisations. All of our discussions will consider averages over different realisations.

The correlation coefficients $\langle f(n_1)f(n_2)\rangle - \langle f(n_1)\rangle\langle f(n_2)\rangle$ are, in general, dependent upon n_1 and n_2. The set of these correlation coefficients is regarded as being a function of n_1 and n_2, called the *correlation function* $C(n_1, n_2)$:

$$C(n_1, n_2) = \langle f(n_1)f(n_2)\rangle - \langle f(n_1)\rangle\langle f(n_2)\rangle. \tag{2.2}$$

The correlation coefficient was introduced in Subsection 1.2.2.

The correlation function is positive if the values of $f(n_2) - \langle f(n_2)\rangle$ tend to have the same sign as $f(n_1) - \langle f(n_1)\rangle$, and negative if they tend to have opposite signs. The correlation function is zero if the values of $f(n_1)$ and $f(n_2)$ are independent. When we investigate the statistics of the random walk in Section 2.4, it will become clear that it is very useful to know the correlation function of a random function.

Recall the discussion in Subsection 1.2.2.

This was shown in the solution to Exercise 1.21.

The task of calculating mean values and correlation functions is greatly simplified if all of the values $f(n)$ are independent, as is the case for the coin-tossing function. These statistics will now be calculated for the coin-tossing function $f(n)$, which takes values ± 1. Recall that the mean value of a random variable is the sum of the possible values of the variable, with each term in the sum multiplied by the probability that this value occurs. In this case, the values ± 1 occur with probability $P(\pm 1) = \frac{1}{2}$, so the mean value of the coin-tossing function is

See equation (1.28).

$$\langle f(n)\rangle = (+1)\,P(+1) + (-1)\,P(-1) = \tfrac{1}{2} - \tfrac{1}{2} = 0. \tag{2.3}$$

Now let us consider the correlation function for the coin-tossing function. Because $\langle f(n)\rangle = 0$ for all n, the correlation function is given by $C(n_1, n_2) = \langle f(n_1)f(n_2)\rangle$, which is the mean of the function evaluated at n_1 multiplied by the function evaluated at n_2. We can see immediately that $\langle f(n_1)f(n_2)\rangle = 0$ when $n_1 \neq n_2$, because the values $f(n_1)$ and $f(n_2)$ are independent. Also, we see that $\langle f(n_1)f(n_2)\rangle = 1$ when $n_1 = n_2$, because $f^2(n)$ is always equal to 1. It follows that

Recall the result of Exercise 1.21: the solution is given for the case of continuous distributions, but the same reasoning applies to the discrete case.

$$\langle f(n_1)f(n_2)\rangle = \begin{cases} 0, & \text{for } n_1 \neq n_2, \\ 1, & \text{for } n_1 = n_2, \end{cases} = \delta_{n_1, n_2} \tag{2.4}$$

(where δ_{n_1, n_2} is the Kronecker delta symbol).

> We can confirm equation (2.4) by a direct calculation of this correlation coefficient using equation (1.41). This requires the joint probability for the function at values n_1 and n_2, which will be defined as follows: $P(s_1, s_2, n_1, n_2)$ is the probability that $f(n_1)$ takes the value s_1 and $f(n_2)$ takes the value s_2. The values of the function at n_1 and n_2 are independent (when $n_1 \neq n_2$), so (according to equation (1.8)) their probabilities multiply. Provided that $n_1 \neq n_2$, and writing $P(s, n)$ for the probability for $f(n) = s$, we have $P(s_1, s_2, n_1, n_2) = P(s_1, n_1)P(s_2, n_2) = \frac{1}{2} \times \frac{1}{2} = \frac{1}{4}$, for any combination of s_1 and s_2. When $n_1 = n_2$, s_1 and s_2 are the same variable, so $P(s_1, s_2, n, n) = 0$ if $s_1 \neq s_2$ (because it is impossible for $f(n)$ to take two different values for a given realisation), whereas $P(s_1, s_2, n, n) = \frac{1}{2}$ if $s_1 = s_2$, because s_1 can be either $+1$ or -1 with equal probability $\frac{1}{2}$. Using these results in equation (1.41), it follows that
>
> $$\begin{aligned}\langle f(n_1)f(n_2)\rangle &= \sum_{s_1 = \pm 1}\sum_{s_2 = \pm 1} P(s_1, s_2, n_1, n_2)\, s_1 s_2 \\ &= (+1)(+1)P(+1, +1, n_1, n_2) + (-1)(-1)P(-1, -1, n_1, n_2) \\ &\quad + (+1)(-1)P(+1, -1, n_1, n_2) + (-1)(+1)P(-1, +1, n_1, n_2) \\ &= \begin{cases} \tfrac{1}{4} + \tfrac{1}{4} - \tfrac{1}{4} - \tfrac{1}{4}, & \text{for } n_1 \neq n_2, \\ \tfrac{1}{2} + \tfrac{1}{2} - 0 - 0, & \text{for } n_1 = n_2, \end{cases} \\ &= \delta_{n_1, n_2}.\end{aligned} \tag{2.5}$$

2.2 An elementary random function

Example 2.1

If a coin has a 'bias', such that the probability of a toss falling heads up is $p \neq \frac{1}{2}$, calculate $\langle f(n) \rangle$, $\langle f(n_1) f(n_2) \rangle$ and $C(n_1, n_2)$ in terms of p.

Solution

The function $f(n)$ takes the value $+1$ if the coin falls heads up, and -1 if the coin falls tails up. If the probability of heads is p, then the probability of tails is $1 - p$. So we have

$$\langle f(n) \rangle = (+1)p + (-1)(1-p) = 2p - 1. \tag{2.6}$$

As the trials are independent, when $n_1 \neq n_2$ we have

$$P(+1, +1, n_1, n_2) = P(+1, n_1)\, P(+1, n_2) = p^2,$$
$$P(-1, -1, n_1, n_2) = P(-1, n_1)\, P(-1, n_2) = (1-p)^2,$$
$$P(+1, -1, n_1, n_2) = P(+1, n_1)\, P(-1, n_2) = p(1-p)$$
$$= P(-1, +1, n_1, n_2). \tag{2.7}$$

If $n_1 = n_2 = n$ (say), then

$$P(\pm 1, \mp 1, n, n) = 0,$$
$$P(+1, +1, n, n) = p,$$
$$P(-1, -1, n, n) = 1 - p. \tag{2.8}$$

For $n_1 \neq n_2$,

$$\langle f(n_1) f(n_2) \rangle = (+1)(+1)p^2 + (-1)(-1)(1-p)^2$$
$$+ (+1)(-1)p(1-p) + (-1)(+1)(1-p)p$$
$$= 4p^2 - 4p + 1$$
$$= (2p - 1)^2 \tag{2.9}$$

(as must be the case, since $C(n_1, n_2) = \langle f(n_1) f(n_2) \rangle - \langle f(n_1) \rangle \langle f(n_2) \rangle = 0$ when $n_1 \neq n_2$). As expected, this vanishes when $p = \frac{1}{2}$. When $n_1 = n_2 = n$, we have

$$\langle [f(n)]^2 \rangle = (+1)^2 p + (-1)^2 (1-p) = 1. \tag{2.10}$$

The results for $\langle f(n_1) f(n_2) \rangle$ can be summarised in a single expression with the help of the Kronecker delta symbol:

$$\langle f(n_1) f(n_2) \rangle = 1 + 4p(p-1)(1 - \delta_{n_1, n_2}). \tag{2.11}$$

The correlation function must equal zero when $n_1 \neq n_2$, because $f(n_1)$ and $f(n_2)$ are independent:

$$C(n_1, n_2) = \langle f(n_1) f(n_2) \rangle - \langle f(n_1) \rangle \langle f(n_2) \rangle$$
$$= 1 + 4p(p-1)(1 - \delta_{n_1, n_2}) - (2p - 1)^2$$
$$= 4p(1-p)\delta_{n_1, n_2}. \tag{2.12}$$

As expected, the results of the discussion preceding this example are recovered when $p = \frac{1}{2}$. ∎

Exercise 2.1

You can generate a discrete random function $f(n)$ by throwing a six-sided die: pick integers n in sequence, and for each n define $f(n)$ to be the number uppermost on throwing the die. What are the values of $\langle f(n) \rangle$, $\langle f(n_1)f(n_2) \rangle$ and the correlation function $C(n_1, n_2)$?

The coin-tossing function is not very useful in its own right. Its principal use is to generate other types of random function, which can be used to model interesting phenomena. In the next section we shall introduce the discrete random walk, which can be developed as a model for the processes illustrated in Figures 0.6 and 0.7. We shall concentrate on discussing random walks, because of their close connection with diffusion processes. At the end of the chapter, in Section 2.8, we introduce another type of random function which can be used to model smoothly varying random systems, such as the temperature series of Figure 0.3 or the model ocean waves shown in Figure 0.5.

2.3 Random walks

2.3.1 Definition and mathematical description

A random walk is a process in which a particle with position X takes a succession of randomly chosen steps at a sequence of times labelled by a number T. The position $X(T)$ is therefore a particular type of random function. We shall use the term 'random walk' for both individual realisations and the process generating these random functions. In most applications, X represents a physical position, but in some applications it can be another type of variable, such as the price of a commodity.

We start by considering the simplest case, which we call the *simple random walk*, which is constructed by the following procedure. Place the particle at the origin, so that at time $T = 0$ its position is $X = 0$. At each step, increase T by 1, and toss a coin. If it falls heads up, increase X by 1; but decrease it by 1 if the coin falls tails up. This process is illustrated in Figure 2.2. Here, T takes non-negative integer values, and X takes integer values.

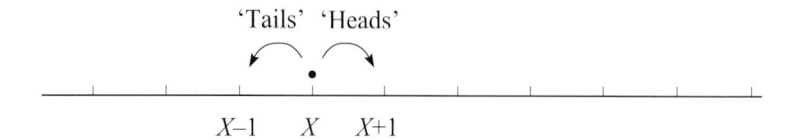

Figure 2.2 Illustrating the process for generating a simple random walk: the particle is moved one step to the right if the coin falls heads up, and one step to the left if it falls tails up

Two realisations of this random walk are plotted in Figure 2.3. Note that X is even when T is even, and X is odd when T is odd.

2.3 Random walks

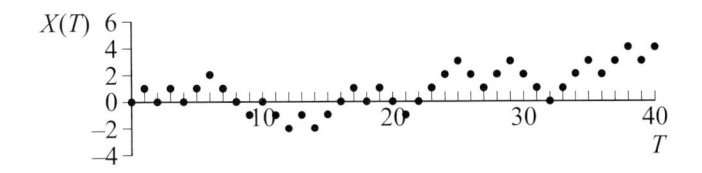

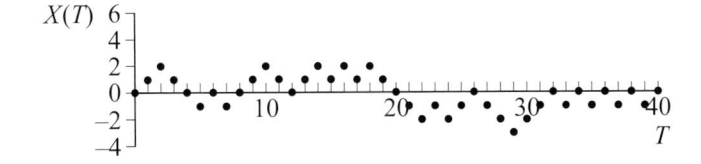

Figure 2.3 Two realisations of the simple random walk

Exercise 2.2

Generate two different simple random walks, each one the result of tossing a coin $T = 10$ times, and plot them in the same fashion as in Figure 2.3. (No solution is given, because there are many different possible outcomes.)

Random walks are excellent models for many phenomena in the physical world, and are also useful models in many other contexts. Some examples will be given shortly, but first we consider how to describe random walk processes mathematically. The simple random walk described above can be constructed from the coin-tossing function $f(n)$ described in Section 2.2. The steps of the variable X, which randomly takes values ± 1 with equal probability, can be taken to be the values of a realisation of the coin-tossing function $f(n)$, so that $X(T)$ is the sum of these values:

$$X(T) = \sum_{n=1}^{T} f(n), \quad \text{where } T = 1, 2, \ldots \text{ and } X(0) = 0. \tag{2.13}$$

It is instructive to express the simple random walk using the recurrence relation

$$X(T) = X(T-1) + f(T), \tag{2.14}$$

which describes more directly how the random walk was generated, stating that the new displacement at time T is the old displacement, $X(T-1)$, at the immediately preceding time, $T-1$, plus a random number, $f(T)$, which takes values ± 1 with equal probability.

Example 2.2

Show that equation (2.13) follows directly from equation (2.14) by repeated application of the recurrence relation.

Solution

From equation (2.14),

$$\begin{aligned}X(T) &= X(T-1) + f(T) \\ &= X(T-2) + f(T-1) + f(T) \\ &= X(0) + \sum_{n=1}^{T} f(n) \\ &= \sum_{n=1}^{T} f(n),\end{aligned} \quad (2.15)$$

where the second line follows by applying the recurrence relation (2.14) to $X(T-1)$ to obtain $X(T-2)$, and the third line comes from repeating this procedure T times. Finally, the last line is a consequence of $X(0) = 0$. ∎

The random walk displacement $X(T)$ at time T is a random variable which can be described by the probability $P(X,T)$ that it equals X after T steps, or by means of its statistics, such as the value of $\langle X^2 \rangle$ at step T. The statistics give a very informative description of a random walk. Section 2.4 will show how these can be calculated directly from the correlation function of the coin-tossing function, without first calculating the probability $P(X,T)$.

2.3.2 Some examples of random walks

The random walk is a very important concept, and has a very wide range of applications in the physical sciences and applied probability. Some of these are described briefly below.

- A very important application of random walks is in the understanding of the phenomenon known as *Brownian motion*. This is the apparently random 'jiggling' motion of small particles of pollen or dust suspended in still water. The motion is not visible to the naked eye, but can be seen using a microscope. It is interpreted as being due to the particle being jostled by the random motion of the molecules making up the liquid (which are too small to be visible even using a microscope). Brownian motion was illustrated in Figure 0.6. It is modelled by a random walk in which the motion occurs in two or three dimensions, and in which both the random displacements and the time intervals between them are very small. Random walks in two or three dimensions and the limiting case of small steps will both be considered in Chapter 6.

  > Brownian motion is named after Robert Brown, a Scottish botanist, who (in 1827) noticed agitated motion of pollen grains in water using a microscope. It would eventually provide important evidence to support the atomic theory of matter. At first, it was believed that the motion of the pollen grains was a result of them being 'alive'. Only after extensive further experiments was it accepted that 'obviously' inanimate materials, such as powdered metals, show the same effect. Even as late as 1865, it was believed that the process of grinding materials might create 'active molecules', which would eventually cease moving. Observations of Brownian motion of particles in sealed containers over a period of one year helped to discredit that theory. By the late nineteenth century, the correct explanation was understood, but not universally accepted.

2.3 Random walks

The reality of atoms continued to be questioned by prominent scientists such as Ernst Mach until the end of the nineteenth century. However, a quantitative analysis of Brownian motion by Josiah Willard Gibbs and Albert Einstein (who worked independently), followed by experimental work by Jean Perrin confirming their theory, showed how evidence from Brownian motion could fit into a consistent atomic theory of matter. Perrin won the 1926 Nobel prize in Physics for his work on sedimentation equilibrium, a phenomenon which is explained by Einstein's theory for Brownian motion.

- The macroscopic process of *diffusion* is a consequence of random motions of molecules. An example of such a random trajectory is illustrated in Figure 2.4: the 'black' molecule follows an apparently random path due to collisions with the 'white' molecules. In liquids and gases it is believed that the molecules all have random trajectories of this type, which are modelled by a three-dimensional extension of the random walk, of the type considered in Chapter 6.

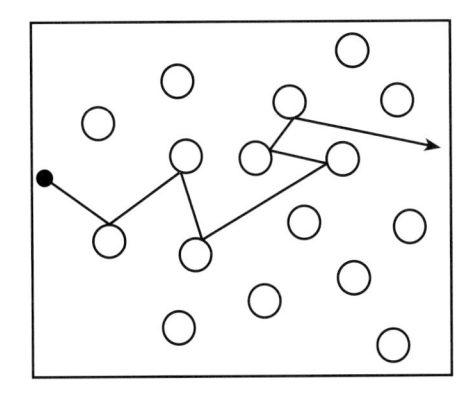

Figure 2.4 Schematic illustration of the highly irregular motion of molecules in a gas: the trajectory of the 'black' atom colliding with the 'white' atoms can be modelled by a random walk in two dimensions

- Apparently random motion (sometimes called chaotic motion) of particles is commonly encountered in many other physical situations. In many cases this motion can be modelled by a random walk. At the smallest scale, the conduction of electricity in metals is explained by models in which electrons (sub-atomic particles that carry electrical current) follow random walks. On the largest scale, the motion of stars through galaxies is sometimes modelled by random walks, in which the steps (resulting from near collisions with other stars) are separated by millions of years.

- Random walks also occur frequently in applied probability. Their earliest appearance was in problems involving games of chance: a gambler's fortune takes random positive or negative steps according to the outcome of a sequence of bets. Such problems are complicated by the fact that the gambler's fortune cannot become negative. They will not be pursued here, because our principal interest is in the connection with diffusion processes.

- Another application of random walks in probability theory is modelling changes in the prices of items traded in financial markets, such as commodities or stocks and shares. A large number of apparently unpredictable decisions by investors to buy and sell causes erratic changes of the market price, such as those illustrated in Figure 0.7.

2.4 Statistics of random walks

The random walk can be characterised by writing down the probability $P(X,T)$ of being at position X after T steps. This is the most complete description, and will be considered shortly. However, it is also very instructive to examine some statistics of $X(T)$: we shall consider the moments $\langle X(T) \rangle$ and $\langle X^2(T) \rangle$. These are readily obtained from equation (2.13), with the help of equations (2.3) and (2.4). The mean value of the displacement of the simple random walk, equation (2.13), is

$$\langle X(T) \rangle = \left\langle \sum_{n=1}^{T} f(n) \right\rangle = \sum_{n=1}^{T} \langle f(n) \rangle = 0. \tag{2.16}$$

(This follows immediately from equations (1.44) and (2.3).) Equation (2.16) tells us that if we generate many realisations of the discrete random walk, and average the displacement X after time T, this average will approach zero as we include more and more realisations. Equation (2.16) reflects the fact that the displacement is equally likely to be in the direction of decreasing X as increasing X.

The square of the displacement is never negative, and therefore gives a more useful indication of the distance X by which the particle is expected to have moved after T steps. Using equations (1.44) and (2.4) gives

$$\langle X^2(T) \rangle = \left\langle \sum_{n_1=1}^{T} f(n_1) \sum_{n_2=1}^{T} f(n_2) \right\rangle$$

$$= \left\langle \sum_{n_1=1}^{T} \sum_{n_2=1}^{T} f(n_1) f(n_2) \right\rangle$$

$$= \sum_{n_1=1}^{T} \sum_{n_2=1}^{T} \langle f(n_1) f(n_2) \rangle$$

$$= \sum_{n_1=1}^{T} \sum_{n_2=1}^{T} \delta_{n_1, n_2}$$

$$= \sum_{n=1}^{T} 1 = T. \tag{2.17}$$

From these results we see that $\mathrm{Var}(X) = \langle X^2 \rangle - (\langle X \rangle)^2 = T$.

More generally, the simple relationship

$$\mathrm{Var}(X) = 2DT \tag{2.18}$$

(where D is a constant called the *diffusion coefficient*) is a characteristic property of random walks. In this case the constant $2D$ is unity, but later we consider cases where it may take other values. It indicates that the typical distance from the starting point (which could be defined as $\sqrt{\mathrm{Var}(X)}$, a quantity which is always positive) is proportional to the square root of the number of steps T.

Exercise 2.3

If a particle, which moves along a line by means of a random walk, has been randomly displaced by a distance 1 mm after 1 s, how far is it likely to have moved after one week? Compare this with the distance moved after one week if the motion is at a constant velocity.

Random walks may have a *drift*, so that the particle moves with a *drift velocity* as well as making random displacements.

Example 2.3

A *random walk with drift* can be obtained as follows. At each step, the particle moves by $v + 1$ with probability $\frac{1}{2}$ or by $v - 1$ with probability $\frac{1}{2}$. Calculate $\langle X \rangle$ and $\langle X^2 \rangle$ after T steps, and show that $\text{Var}(X) = T$. Show that $\langle X \rangle = vT$, so that v can be thought of as a velocity, called the *drift velocity*.

Solution

Here the displacements are a random function $f(n)$ which takes value $v + 1$ or $v - 1$, both with probability $\frac{1}{2}$. The mean of $f(n)$ is

$$\langle f(n) \rangle = \tfrac{1}{2}(v+1) + \tfrac{1}{2}(v-1) = v. \tag{2.19}$$

Because values of $f(n)$ for different values of n are independent, for $n_1 \neq n_2$ we have $C(n_1, n_2) = 0$, so

$$\langle f(n_1) f(n_2) \rangle = \langle f(n_1) \rangle \langle f(n_2) \rangle = v^2 \quad \text{for } n_1 \neq n_2. \tag{2.20}$$

Also,

$$\langle [f(n)]^2 \rangle = \tfrac{1}{2}(v+1)^2 + \tfrac{1}{2}(v-1)^2 = v^2 + 1. \tag{2.21}$$

The random walk is a sum of values of the random function $f(n)$:

$$X(T) = \sum_{n=1}^{T} f(n). \tag{2.22}$$

The mean and second moment of $X(T)$ are

$$\langle X(T) \rangle = \sum_{n=1}^{T} \langle f(n) \rangle = vT \tag{2.23}$$

and

$$\langle [X(T)]^2 \rangle = \sum_{n_1=1}^{T} \sum_{n_2=1}^{T} \langle f(n_1) f(n_2) \rangle$$
$$= T(v^2 + 1) + (T^2 - T)v^2 = T^2 v^2 + T, \tag{2.24}$$

because there are T terms where $n_1 = n_2$ and $T^2 - T$ terms where $n_1 \neq n_2$. From these we find $\text{Var}(X) = \langle X^2 \rangle - (\langle X \rangle)^2 = T$. Note that the variance is proportional to T, in agreement with equation (2.18). ∎

Exercise 2.4

Consider a process in which, at each step, a particle is displaced by $+1$ with probability p or by -1 with probability $1-p$. Calculate $\langle X \rangle$, $\langle X^2 \rangle$ and $\mathrm{Var}(X)$ after T steps. In view of the discussion in Example 2.3 and the discussion following equation (2.18), does this process also describe a random walk with drift?

Some harder exercises on statistics of random walks can be found at the end of this chapter.

2.5 Probability distribution of a random walk

The aim of this section is to formulate and solve an equation for the probability $P(X,T)$ of reaching X after T steps, in the case of the simple random walk illustrated in Figure 2.2. If we generate M realisations of the simple random walk with T steps, and count the number M_X of these realisations for which the particle finishes at position X, we expect that in the limit as $M \to \infty$, the ratio M_X/M approaches a limit $P(X,T)$. We shall now calculate $P(X,T)$ using the rules of probability theory described in Chapter 1.

The particle can reach position X at step T either from being at position $X+1$ at step $T-1$ and throwing tails (so the particle takes a step back), or from being at position $X-1$ at step $T-1$ and throwing heads (so the particle takes a step forward) – see Figure 2.5. The probability of reaching X the first way is

$$P_{\text{back}} = P(-1) \times P(X+1, T-1), \tag{2.25}$$

where $P(-1) = \frac{1}{2}$ is the probability of taking a decreasing step. (Note that these probabilities multiply because the probability of taking a step to the left or right is independent of the current position – recall equation (1.8).) Similarly, the second route contributes the probability

$$P_{\text{forward}} = P(+1) \times P(X-1, T-1), \tag{2.26}$$

where $P(+1) = \frac{1}{2}$ is the probability of taking an increasing step.

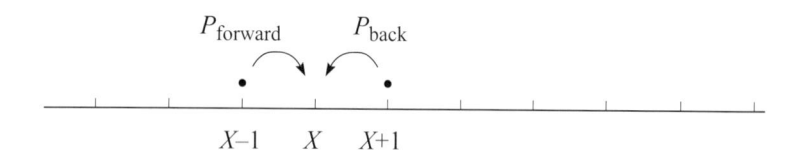

Figure 2.5 The random walker can arrive at position X from two possible previous locations. The probability $P(X,T)$ of reaching X at time T is the sum of the probabilities for these steps.

Since arriving at X by taking a step forward and reaching X by a step back are mutually exclusive events, the probability of being at X at time $T > 0$ is the sum of these terms (recall equation (1.6)). Thus we have $P(X,T) = P_{\text{forward}} + P_{\text{back}}$, or

$$P(X,T) = \tfrac{1}{2}\left[P(X-1,T-1) + P(X+1,T-1)\right]. \tag{2.27}$$

2.5 Probability distribution of a random walk

This is a *recurrence relation*, giving the probability at step T in terms of the values at the preceding step, $T-1$. The values of $P(X,T)$ may be obtained for any positive integer T, using the initial data $P(0,0) = 1$ and $P(X \neq 0, 0) = 0$.

Table 2.1 gives the values of $P(X,T)$ up to $T=4$, obtained by a direct calculation of all of the $P(X,1)$, then all of the $P(X,2)$, and so on, using the recurrence formula (2.27).

The initial conditions can be expressed concisely in terms of the Kronecker delta symbol: $P(X,0) = \delta_{X,0}$.

Table 2.1 Probabilities for the simple random walk up to $T = 4$

$P(X,T)$					$X(T)$				
T	-4	-3	-2	-1	0	1	2	3	4
0	0	0	0	0	1	0	0	0	0
1	0	0	0	$\frac{1}{2}$	0	$\frac{1}{2}$	0	0	0
2	0	0	$\frac{1}{4}$	0	$\frac{2}{4}$	0	$\frac{1}{4}$	0	0
3	0	$\frac{1}{8}$	0	$\frac{3}{8}$	0	$\frac{3}{8}$	0	$\frac{1}{8}$	0
4	$\frac{1}{16}$	0	$\frac{4}{16}$	0	$\frac{6}{16}$	0	$\frac{4}{16}$	0	$\frac{1}{16}$

Exercise 2.5

Use the recurrence relation (2.27) to determine the probabilities in the next two rows of Table 2.1, for $T = 5$ and $T = 6$.

We can solve equation (2.27) by investigating a connection with *Pascal's triangle*, which is illustrated in Figure 2.6.

```
                    1
                1       1
            1       2       1
        1       3       3       1
    1       4       6       4       1
1       5      10      10       5       1
```

Figure 2.6 The first six rows of Pascal's triangle

Pascal's triangle (said to have been 'discovered' by Blaise Pascal, 1623–62, although already known and extensively used by Arab, Chinese and Indian mathematicians at least five centuries earlier) is a table of integers in a triangular format. Each entry is the sum of the two numbers from the row above immediately to the right and to the left. If there is no number entered in one of these positions, as for the first and last entries of each row, zero is added. For example, in row six at position three, the entry is obtained by adding entries two and three of row five, giving $4 + 6 = 10$.

Pascal's triangle was originally devised as a construction to obtain the *binomial coefficients* C^n_r. When the *binomial* $(x+y)^n$ is expanded, terms proportional to $x^r y^{n-r}$ are obtained, with r taking values from 0 to n. The coefficient of each such term is the number C^n_r, which is also the $(r+1)$th number in the $(n+1)$th row of Pascal's triangle, starting from $C^n_0 = 1$.

The formula for C^n_r will be given shortly.

The binomial coefficient C^n_r is also the number of ways of selecting r choices from n objects, if the order in which the choice is made is irrelevant.

Exercise 2.6

Determine the numbers in the next row of Pascal's triangle. Also, expand $(1+x)^6$, and confirm that the coefficients of x^r are given by the numbers that you have just calculated.

Table 2.1 should be contrasted with the first five rows of Pascal's triangle. It can be seen that the probabilities for any given value of T are equal to corresponding entries in the $(T+1)$th row of Pascal's triangle, divided by 2^T.

One way to understand the connection with Pascal's triangle is to use the interpretation of its elements C_r^n as the number of ways of selecting r choices from n objects. After T steps, the simple random walk has made N positive steps and $T-N$ negative steps. Each possible path for the walk involves choosing T steps independently, each with probability $\frac{1}{2}$, so the choice of any particular realisation has probability $(\frac{1}{2})^T$. The probability $P(X,T)$ of choosing any of the C_N^T random walks of N positive steps from a total of T is

$$P(X,T) = \frac{1}{2^T} C_N^T, \qquad (2.28)$$

where X and T are such that there are exactly N positive steps. It is more natural to express the right-hand side of equation (2.28) in terms of X and T rather than N and T. This can be done by relating the number of positive steps N to the position X: we have $X = N \times (+1) + (T-N) \times (-1)$, so $X = 2N - T$. Rearranging gives $N = (X+T)/2$. Expressing equation (2.28) in terms of X, we find that equation (2.27) is expected to have a solution of the form

$$P(X,T) = \frac{1}{2^T} C_{(X+T)/2}^T. \qquad (2.29)$$

Note that this solution applies only when X and T have the same *parity* (i.e. both even or both odd). As can be seen from the expression $X = 2N - T$, it is impossible for the parity of the position X reached by the random walk to differ from that of the step index T, so when the parities of X and T differ, $P(X,T) = 0$ (see Table 2.1).

Equation (2.29) is the required solution of equation (2.27) when X and T have the same parity. In order to make use of this solution, it is necessary to have a formula for the binomial coefficients. The coefficient C_r^n in the $(r+1)$th location along the $(n+1)$th row of Pascal's triangle (with r running from 0 to n) is given by

$$C_r^n = \frac{n!}{r!(n-r)!} \qquad (2.30)$$

where $n!$ is the *factorial* of n, given by $n! = n \times (n-1) \times \cdots \times 3 \times 2 \times 1$, and (by convention) $0! = 1$.

The arguments presented in the preceding two paragraphs deduced equation (2.29) from known facts about the binomial coefficients, without giving a proof. The following exercise verifies that equation (2.30) gives the entries in Pascal's triangle, and that solution (2.29) satisfies equation (2.27).

Exercise 2.7

The coefficients in Pascal's triangle satisfy the recurrence relation

$$C_r^n = C_r^{n-1} + C_{r-1}^{n-1}, \tag{2.31}$$

for $n \geq 1$, where C_r^n is the $(r+1)$th coefficient in the $(n+1)$th row, and coefficients with $r \leq -1$ or $r \geq n+1$ are taken to be zero. Note the close similarity of this relation to equation (2.27). Show that equation (2.30) satisfies this relation, and also that solution (2.29) satisfies equation (2.27).

Exercise 2.8

Write down the equation analogous to (2.27) which applies to the model discussed in Exercise 2.4.

Equation (2.29) is an exact formula, but is not particularly instructive, because the factorial functions are not easy to use in subsequent calculations. The next section gives a simple and very useful approximation for $P(X, T)$.

2.6 An approximate form for the probability distribution

Figure 2.7 shows the values of the probabilities $P(X, T)$ given by equation (2.29) plotted as a function of X for two fixed values of T, compared with an approximate form $P_{\text{app}}(X, T)$ given by the expression

$$P_{\text{app}}(X, T) = \frac{2}{\sqrt{2\pi T}} \exp(-X^2/2T). \tag{2.32}$$

For clarity, the figure plots $P_{\text{app}}(X, T)$ for $T = 3$ and $T = 11$ as two continuous curves despite the fact that they represent probabilities at integer values of X, not probability densities. The exact values of $P(X, T)$ are shown as dots. This figure shows that equation (2.32) is a very good approximation, and its accuracy is seen to improve as T increases. It should be understood that this approximate distribution applies only to values of X with the same parity (odd or even) as T, and that the probability must be taken to be zero for the opposite parity (as shown by the dots on the X-axis in Figure 2.7). The approximate probability (2.32) is a Gaussian function, with a variance that is proportional to T.

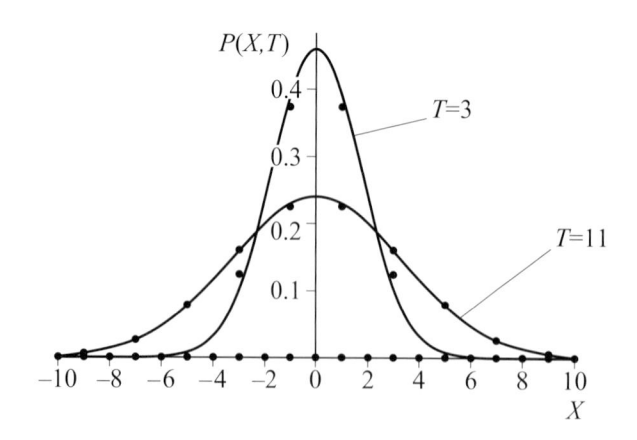

Figure 2.7 Comparison of exact probabilities for a random walk (dots) with a Gaussian approximation (continuous curve), for the cases $T = 3$ and $T = 11$

Exercise 2.9

Hard exercise

Estimate $\langle X^2 \rangle$ for the approximate probability (2.32), by approximating the sum over values of X by an integral. Verify that the result is identical to that for the exact probability, given by equation (2.17).

We shall now consider why solution (2.29) can be approximated by the Gaussian function (2.32), using an approximation for the logarithm of the factorial function $N!$ known as *Stirling's formula*:

$$\ln(N!) = N \ln(N) - N + \tfrac{1}{2} \ln(2\pi N) + O(1/N). \tag{2.33}$$

The final term indicates that N times the magnitude of the error is bounded in the limit as $N \to \infty$.

Exercise 2.10

Using an electronic calculator to evaluate Stirling's formula, estimate $6!$, $8!$ and $10!$. Compare these estimates with the exact values.

Confirm that the relative error decreases as N increases. (If x is an approximation to x_0, the relative error is defined as $|x_0 - x|/x_0$.)

Stirling's formula is a far from obvious result; the following remarks give some insight into the form of equation (2.33).

Let us obtain upper and lower bounds for $\ln(n!)$ by using the function

$$L(x) = \ln\left[\mathrm{Int}(x+1)\right], \tag{2.34}$$

where $\mathrm{Int}(x)$ denotes the *integer part* of x. Note that

$$\ln(n!) = \sum_{j=1}^{n} \ln(j) = \int_1^n dx\, L(x), \tag{2.35}$$

For example, $\mathrm{Int}(23.764) = 23$, and we shall consider only $x > 0$.

because the integrand is constant over a succession of intervals of unit length (see Figure 2.8). Also note that since $x < \mathrm{Int}(x+1) \leq x+1$,

$$\ln(x) < L(x) \leq \ln(x+1) \tag{2.36}$$

(as also shown in Figure 2.8). Finally, integrating this inequality from 1 to n, we conclude that

$$n \ln(n) - n + 1 \leq \ln(n!) \leq (n+1)\ln(n+1) - (n+1) - 2\ln(2) + 2. \tag{2.37}$$

This is consistent with Stirling's formula, and gives some insight into the form of the two leading terms.

2.6 An approximate form for the probability distribution

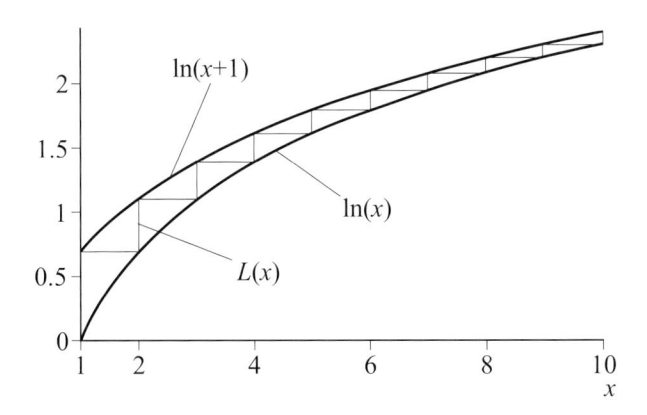

Figure 2.8 The function $L(x) = \ln[\text{Int}(x+1)]$ is bounded above by $\ln(x+1)$ and below by $\ln(x)$

Exercise 2.11

The step from equation (2.36) to equation (2.37) requires the integral of $\ln(x)$. What is this integral?

Now the expression (2.29) for $P(X, T)$ will be approximated using Stirling's formula. First write it in terms of factorial functions, using equation (2.30):

$$P(X,T) = \frac{1}{2^T} C^T_{(X+T)/2} = \frac{T!}{2^T \left(\frac{1}{2}(T+X)\right)! \left(\frac{1}{2}(T-X)\right)!} \quad (2.38)$$

(where $X + T$ is even). Now take the natural logarithm of this equation, using the rule $\ln(a/b) = \ln(a) - \ln(b)$, and apply Stirling's approximation to each factorial:

$$\ln[P(X,T)] = \ln(T!) - \ln\left[\left(\tfrac{1}{2}(T+X)\right)!\right] - \ln\left[\left(\tfrac{1}{2}(T-X)\right)!\right] - T\ln(2)$$

$$\simeq T\ln(T) - \left(\frac{T+X}{2}\right)\ln\left(\frac{T+X}{2}\right)$$

$$- \left(\frac{T-X}{2}\right)\ln\left(\frac{T-X}{2}\right) - T\ln(2)$$

$$+ \tfrac{1}{2}\ln\left[\frac{2T}{\pi(T+X)(T-X)}\right]. \quad (2.39)$$

Exercise 2.12

Check that you can obtain equation (2.39) by applying formula (2.33) to equation (2.38) and dropping the error terms.

Equation (2.39) can be simplified, using an approximation which is valid when $|X/T| \ll 1$. The following Taylor series will be useful for expanding the terms involving logarithms:

$$\ln(1+\lambda) = \lambda - \frac{\lambda^2}{2} + \frac{\lambda^3}{3} + O(\lambda^4). \tag{2.40}$$

Writing $\epsilon = X/T$ and using equation (2.40), we find after a series of steps that equation (2.39) gives the simple approximation

$$\ln(P) \simeq -\frac{T\epsilon^2}{2} + \tfrac{1}{2}\ln\left(\frac{2}{\pi T}\right). \tag{2.41}$$

The steps leading to this expression are the subject of an end-of-chapter exercise.

Exponentiating, and substituting $\epsilon = X/T$, gives

$$P(X,T) \simeq \frac{2}{\sqrt{2\pi T}} \exp(-X^2/2T), \tag{2.42}$$

which is the Gaussian approximation introduced as equation (2.32). Thus $P(X,T)$ is approximated by a Gaussian distribution when $T \gg 1$ and $X/T \ll 1$. This approximation is very useful because the Gaussian function has more convenient properties than the factorial functions appearing in equation (2.29).

The end-of-chapter exercises include a structured exercise (Exercise 2.18) to help you to fill in the gaps going from (2.39) to (2.41). The calculation is lengthy, and is a good test of your ability to do algebra, but it contains no important additional ideas.

2.7 Relationship with the diffusion equation

This section will present an alternative approach to explaining why the probability distribution of a simple random walk is well approximated by a Gaussian distribution. It is based upon noting a relationship between a version of equation (2.27) and a partial differential equation of the form

$$\frac{\partial P}{\partial T} = D \frac{\partial^2 P}{\partial X^2}. \tag{2.43}$$

Equation (2.43) is an important equation of applied mathematics, known as the *diffusion equation*, and the constant D is called the *diffusion coefficient*. In later chapters it will be shown that the connection with the diffusion equation is not coincidental.

See also equation (2.18), where D was also encountered.

To understand how equation (2.43) relates to a random walk, we consider a random walk in which a particle takes steps of length δX (instead of steps of length 1 as before) at times separated by δT, such that there are equal probabilities for jumping to the left or right. Repeating the argument leading to equation (2.27), we obtain

$$P(X, T+\delta T) = \tfrac{1}{2}\left[P(X+\delta X, T) + P(X-\delta X, T)\right]. \tag{2.44}$$

It will be assumed that δT and δX are small. Subtracting $P(X,T)$ from both sides, we have

$$P(X, T+\delta T) - P(X, T)$$
$$= \tfrac{1}{2}\left[P(X+\delta X, T) + P(X-\delta X, T) - 2P(X,T)\right]. \tag{2.45}$$

Now we make a Taylor series expansion of all the terms about (X, T), and find

$$P(X, T + \delta T) - P(X, T)$$
$$= \frac{\partial P}{\partial T}(X,T)\,\delta T + \tfrac{1}{2}\frac{\partial^2 P}{\partial T^2}(X,T)\,(\delta T)^2 + \cdots \qquad (2.46)$$

and

$$\tfrac{1}{2}\left[P(X+\delta X, T) + P(X-\delta X, T) - 2P(X,T)\right]$$
$$= \tfrac{1}{2}\frac{\partial^2 P}{\partial X^2}(X,T)\,(\delta X)^2 + \tfrac{1}{24}\frac{\partial^4 P}{\partial X^4}(X,T)\,(\delta X)^4 + \cdots. \qquad (2.47)$$

If we assume that both δX and δT are sufficiently small that only the leading terms of both these Taylor series need be retained, then on substituting these terms into equation (2.45) we have

$$\frac{\partial P}{\partial T} \simeq \frac{(\delta X)^2}{2\,\delta T} \frac{\partial^2 P}{\partial X^2}, \qquad (2.48)$$

so the recurrence relation (2.44) can be approximated by the diffusion equation (2.43), with diffusion coefficient $D = (\delta X)^2/2\,\delta T$.

Now we note that the Gaussian function $P_{\text{app}}(X,T)$ in equation (2.32) is an exact solution of the diffusion equation. Demonstrating this fact is left to the following exercise.

Exercise 2.13

Verify that $P_{\text{app}}(X,T) = 2\exp(-X^2/2T)/\sqrt{2\pi T}$ is (for $T > 0$) an exact solution of the diffusion equation when $D = \tfrac{1}{2}$.

Earlier (in Section 2.6) the use of the Gaussian function $P_{\text{app}}(X,T)$ was justified by showing that it approximates the exact solution. This second approach has demonstrated that it is also an exact solution of an approximate version of equation (2.44), namely the diffusion equation.

2.8 A random function with correlations

We started this chapter by introducing the coin-tossing function $f(n)$, a random function for which successive values are uncorrelated, satisfying $C(n_1, n_2) = 0$ for $n_1 \neq n_2$. In many situations it is necessary to model random functions with correlations. An example of this would be modelling the sequence of daily temperature records illustrated in Figure 0.3; these are apparently random, but temperatures on successive days are clearly correlated. In this and many other examples, the correlation function is expected to depend only upon the difference in time between observations, so that the correlation function may be written in terms of a function $\mathcal{C}$ of a single variable: $C(n_1, n_2) = \mathcal{C}(n_1 - n_2)$. A more complex example, involving a correlated random function of two variables, is the height of the surface of the ocean on a windy day, illustrated in Figure 0.5. This section will show how a discrete model for such correlated random functions may be constructed, and will examine its correlation properties.

First, we consider a very simple case of a random function with correlations. Let $M > 1$ be an integer, and let $f(n)$ be a realisation of the coin-tossing function discussed in Section 2.2. Now let $g(n)$ be the average of the last M values of $f(n)$, that is,

$$g(n) = \frac{1}{M}[f(n) + f(n-1) + f(n-2) + \cdots + f(n+1-M)]$$
$$= \frac{1}{M}\sum_{j=1}^{M} f(n+1-j). \tag{2.49}$$

The function $g(n)$ is called a *running average* over the last M values of $f(n)$. A realisation of $g(n)$ is plotted in Figure 2.10. The values of $g(n)$ and $g(n+1)$ will be correlated, because they both contain the same random numbers $f(n), f(n-1), \ldots, f(n+2-M)$. However, $g(n)$ and $g(n+20M)$ (say) are not correlated, because the values of $f(n)$ that contribute to $g(n)$ are distinct from those that contribute to $g(n+20M)$. We consider the correlation properties of $g(n)$ in the following example.

Example 2.4

What is the correlation function $C(n_1, n_2)$ for the random function defined by equation (2.49)?

Solution

The correlation function is $C(n_1, n_2) = \langle g(n_1)g(n_2)\rangle - \langle g(n_1)\rangle\langle g(n_2)\rangle$. We see immediately that $\langle g(n)\rangle = 0$ for all n, because $\langle f(m)\rangle = 0$ for all m. The correlation function is therefore

$$C(n_1, n_2) = \langle g(n_1)g(n_2)\rangle$$
$$= \frac{1}{M^2}\sum_{j=1}^{M}\sum_{k=1}^{M}\langle f(n_1+1-j)f(n_2+1-k)\rangle$$
$$= \frac{1}{M^2}\sum_{j=1}^{M}\sum_{k=1}^{M}\delta_{n_1+1-j,n_2+1-k}. \tag{2.50}$$

The final step uses equation (2.4). The double summation therefore counts the number of common elements of the sets $\{n_1, n_1-1, \ldots, n_1+1-M\}$ and $\{n_2, n_2-1, \ldots, n_2+1-M\}$. The elements of these two sets are shown as bold marks in Figure 2.9, and the common elements are indicated by vertical lines. We see that the number of common elements is equal to M when $n_1 = n_2$, and reduces by one every time $|n_1 - n_2|$ increases by one, until it becomes zero when $|n_1 - n_2| = M$. It follows that

$$C(n_1, n_2) = \begin{cases} \frac{M - |n_1 - n_2|}{M^2}, & \text{for } 0 \leq |n_1 - n_2| \leq M, \\ 0, & \text{otherwise.} \end{cases} \tag{2.51}$$

2.8 A random function with correlations

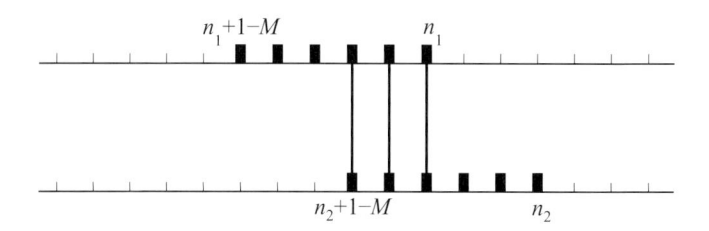

Figure 2.9 The correlations of the running average result from those values of n for which $f(n)$ contributes to both $g(n_1)$ and $g(n_2)$: the values of $f(n)$ contributing to each respective function are indicated by bold marks, and the common values are indicated by vertical lines. Here $M = 6$, $n_2 - n_1 = 3$, and there are $M - |n_1 - n_2| = 3$ common values of n.

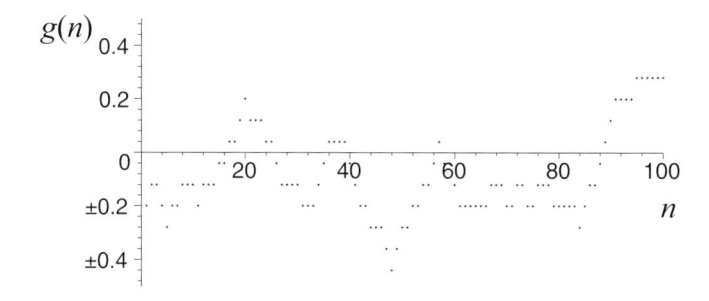

Figure 2.10 A realisation of the correlated random function $g(n)$ defined in equation (2.49) as the running average of M values of a realisation of the coin-tossing function, where $M = 25$ ∎

Thus we have seen that taking a running average over the last M values of the coin-tossing function can be used as a correlated random function. A more general correlated discrete random function can be constructed as follows. Let $w(n)$ be a *weight function* (note that, here, $w(n)$ is never random), which approaches zero rapidly as $|n| \to \infty$, and let $f(n)$ be a realisation of the coin-tossing function introduced in Section 2.2, with a correlation function given by equation (2.4). We define

$$g(n) = \sum_{m=-\infty}^{\infty} w(n-m)f(m), \qquad (2.52)$$

which is a random function (because $f(n)$ is random). The running average in equation (2.49) is a special case of this construction, where $w(n) = 1/M$ for $0 \leq n \leq M - 1$, and $w(n) = 0$ otherwise. Equation (2.52) is a discrete convolution of the random function $f(m)$ and the weight function $w(n)$. In many applications $w(n)$ would be non-zero for all n, for example a Gaussian function, $w(n) = A\exp[-(n/r)^2]$, or an exponential function, $w(n) = A\exp(-|n|/r)$. Normally, it is desired that many values of $f(m)$ make a significant contribution to the sum in equation (2.52), so r is chosen to be large compared to unity.

Discrete convolutions were mentioned in Block I, Chapter 3: see equation (3.47).

Figure 2.11 shows a realisation of the function $g(n)$ defined by equation (2.52) in the case where w is a Gaussian function, $w(n) = \exp[-(n/2)^2]$. The function values are unpredictable, but $g(n_1)$ and $g(n_2)$ are correlated in that they tend to take nearby values, over a range of approximately $|n_1 - n_2| \simeq 2r = 4$. This random function is still defined only for integer values of the argument n, but the function values $g(n)$ are now real numbers, rather than integers. The process of taking the discrete convolution with $w(n)$ 'smoothes out' the highly erratic fluctuations of the coin-tossing function, so although $g(n)$ is still a random function, it has a pleasingly 'smooth' dependence on n.

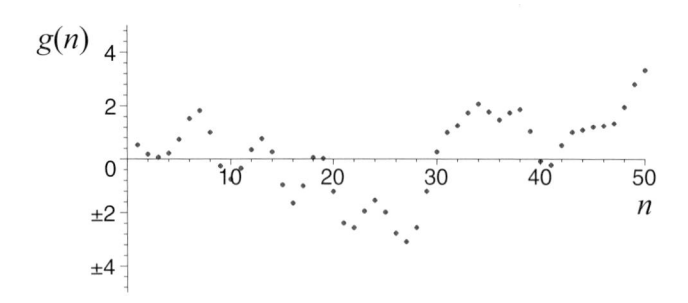

Figure 2.11 A realisation of the correlated random function $g(n)$ defined in equation (2.52). The weight function is a Gaussian, $w(n) = \exp[-(n/2)^2]$.

There are two respects in which equation (2.52) is more useful than equation (2.49) as a model for a correlated random function. First, the form of the correlation function for the running average, equation (2.49), is prescribed, whereas different choices of $w(n)$ enable the correlation function of $g(n)$, given by equation (2.52), to be varied to match different applications. Secondly, for some choices of $w(m)$, the function $g(n)$ takes a continuous range of values, which is more natural when modelling many physical phenomena.

Now we shall develop a theory for the correlation properties of the function $g(n)$ defined in equation (2.52). The approach is to obtain statistics for $g(n)$ in terms of the statistics for the coin-tossing function $f(n)$. This will be done in the following exercise.

Exercise 2.14

For $g(n)$ as given in equation (2.52), show that $\langle g(n) \rangle = 0$. [Hint: Use equation (1.44).]

Show that the correlation function of $g(n_1)$ and $g(n_2)$ is a function of $n_2 - n_1$, given by

$$\mathcal{C}(n_2 - n_1) = \langle g(n_1) g(n_2) \rangle = \sum_{m=-\infty}^{\infty} w(n_1 - n_2 + m)\, w(m). \tag{2.53}$$

[Hint: Start with the definition of the correlation function of g, i.e. $C(n_1, n_2) = \langle g(n_1) g(n_2) \rangle - \langle g(n_1) \rangle \langle g(n_2) \rangle$, and express the function g in terms of the coin-tossing function f using equation (2.52). Now use equation (1.44) to write the average of the double sum as a sum of averages, obtaining a double sum over $\langle f(m_1) f(m_2) \rangle$. Finally, use the expression (2.4) for this correlation, reduce the double sum to a single summation, and make a change of variable.]

Exercise 2.15

Show that the correlation function $\mathcal{C}(n)$ obtained in Exercise 2.14 satisfies

$$\sum_{n=-\infty}^{\infty} \mathcal{C}(n) > 0 \tag{2.54}$$

(unless $w(n)$ is identically zero for all n).

2.9 Summary and discussion

Summary

The important points from this chapter are listed below.

- Random functions are important for modelling many situations. They may be described by their statistical properties: for example, a random function $f(n)$ defined for integer n can be described by its mean $\langle f(n) \rangle$ and correlation function $C(n_1, n_2) = \langle f(n_1)f(n_2)\rangle - \langle f(n_1)\rangle\langle f(n_2)\rangle$. We introduced a very simple random function, the coin-tossing function, with simple statistics $\langle f(n)\rangle = 0$ and $C(n_1, n_2) = \delta_{n_1, n_2}$.

- Other random functions can be obtained from the coin-tossing function $f(n)$ by taking linear combinations of the values $f(n)$. Two examples were considered: the random walk was introduced in Section 2.3, and the correlated random function in Section 2.8. The discrete random walk $X(T)$ is a sum of T values of the coin-tossing function: $X(T) = \sum_{n=1}^{T} f(n)$. The correlated random function $g(n)$ is a discrete convolution of $f(n)$ with a weight function $w(n)$. These two types of random function have different properties and applications.

- A characteristic property of random walks is that the variance of the displacement $\text{Var}(X)$ is proportional to T, implying that the typical spread of the displacement (which could be defined as the standard deviation of X) is proportional to $\sqrt{T}$.

- For large values of T, the probability distribution of a random walk is well approximated by a Gaussian function. Stirling's formula was used to establish this.

- The probability distribution of a discrete random walk satisfies a recurrence relation, which can be approximated by the diffusion equation. This indicates a close relationship between random walks and the phenomenon of diffusion.

Discussion

There are two important issues which are left for discussion in later chapters.

- It was shown that the random walk is closely related to the diffusion equation, which is used to describe the apparently deterministic motion of dissolved materials and of heat. It is important to understand the relationship between the diffusion equation and the random walk. Before this can be done, the diffusion equation itself must be understood, and the next two chapters will introduce this equation, and methods for its solution. Later chapters will show how diffusion, an apparently deterministic macroscopic process, arises from the random microscopic motion of molecules.

- The random function and random walk were defined in this chapter only on a discrete space, the set of integers. For many physical applications, functions defined upon a continuum are required – in one dimension, this is the real line. Most of the properties that have been described above extend in a natural way to functions defined on the real line. Many applications of random functions defined as functions of a continuous variable are related to diffusion processes. It is therefore natural to defer discussion of these until after work on the diffusion equation. Later, in Chapter 6, we shall discuss random walks defined on the real line.

2.10 Outcomes

In addition to being aware of the points listed in Section 2.9, after studying this chapter you should:
- be able to calculate simple statistical properties for discrete random functions which are closely related to those already discussed in this chapter;
- be able to formulate recurrence relations for the probability distributions of random walks, similar in form to equation (2.27);
- be able to use Stirling's formula to approximate expressions involving factorials.

2.11 Further Exercises

Here are two much harder exercises on the statistics of the simple random walk. They are tackled using the approach developed in Section 2.4.

Exercise 2.16

Consider the quantity $\langle X(T_1)X(T_2)\rangle$ for a simple random walk. (This is the mean value of the product of the position $X(T_1)$ at time T_1 and the position of the same realisation $X(T_2)$ at time T_2.) Show that $\langle X(T_1)X(T_2)\rangle = \min(T_1, T_2)$, where

$$\min(a,b) = \begin{cases} a, & a \leq b, \\ b, & a > b. \end{cases} \tag{2.55}$$

Exercise 2.17

Show that for the simple random walk,

This is a very hard exercise.

$$\langle f(n_1)f(n_2)f(n_3)f(n_4)\rangle = \delta_{n_1,n_2}\delta_{n_3,n_4} + \delta_{n_1,n_3}\delta_{n_2,n_4} + \delta_{n_1,n_4}\delta_{n_2,n_3} \\ - 2\delta_{n_1,n_2}\delta_{n_1,n_3}\delta_{n_1,n_4}, \tag{2.56}$$

where $f(n)$ is the coin-tossing function introduced in Section 2.2 which takes the values ± 1. Hence show that for the discrete random walk (2.13), we have

$$\langle X^4(T)\rangle = 3T^2 - 2T. \tag{2.57}$$

Compare this with the result of Exercise 1.28, showing that there is agreement at leading order as $T \to \infty$.

Do not worry if you find the worked solution difficult. The result is instructive, but not essential to understanding other parts of this chapter.

2.11 Further Exercises

The following exercise will guide you through the calculation to obtain equation (2.41) from equation (2.39).

Exercise 2.18

Writing $\epsilon = X/T$ and using the Taylor series (2.40), show that

$$\left(\frac{T+X}{2}\right)\ln\left(\frac{T+X}{2}\right) = \frac{T}{2}(1+\epsilon)\left[\ln\left(\frac{T}{2}\right) + \ln(1+\epsilon)\right]$$
$$= \frac{T}{2}\ln\left(\frac{T}{2}\right) + \frac{T}{2}\left[\ln\left(\frac{T}{2}\right) + 1\right]\epsilon$$
$$+ \frac{T}{4}\epsilon^2 + O(\epsilon^3). \tag{2.58}$$

An analogous expression for the term in equation (2.39) containing $T - X$ is obtained by changing the sign of ϵ. Substitute these expressions into equation (2.39) to obtain

$$\ln[P(X,T)] \simeq \tfrac{1}{2}\ln\left[\frac{4T}{2\pi(T^2 - X^2)}\right] - \frac{T}{2}\epsilon^2$$
$$= \tfrac{1}{2}\ln\left(\frac{4}{2\pi T}\right) - \frac{T}{2}\epsilon^2 - \tfrac{1}{2}\ln(1-\epsilon^2). \tag{2.59}$$

Now use equation (2.40), with λ replaced by $-\epsilon^2$, to show that the final term of equation (2.59) may be approximated by $\epsilon^2/2$. When $T \gg 1$, this term is negligible compared to the term $-T\epsilon^2/2$ of equation (2.59), leading to equation (2.41).

Solutions to Exercises in Chapter 2

Solution 2.1

The possible values of $f(n)$ are $\{1,2,3,4,5,6\}$, each occurring with probability $\frac{1}{6}$. The mean value is

$$\langle f(n) \rangle = \sum_i P_i\, i = (1+2+3+4+5+6)/6 = 21/6 = 7/2.$$

When $n_1 \neq n_2$, the random variables $f(n_1)$ and $f(n_2)$ are independent, so

$$\langle f(n_1) f(n_2) \rangle = (7/2)^2 = 49/4.$$

When $n_1 = n_2$, we have

$$\langle f(n_1) f(n_2) \rangle = \langle [f(n)]^2 \rangle = \sum_i P_i\, i^2 = (1+4+9+16+25+36)/6 = 91/6.$$

The correlation function $C(n_1, n_2)$ is equal to zero when $n_1 \neq n_2$, because $f(n_1)$ and $f(n_2)$ are independent. (This can be checked by calculating $C(n_1, n_2) = \langle f(n_1) f(n_2) \rangle - \langle f(n_1) \rangle \langle f(n_2) \rangle$ directly, from the results above.)

When $n_1 = n_2 = n$, we have

$$C(n,n) = \langle [f(n)]^2 \rangle - [\langle f(n) \rangle]^2 = 91/6 - (7/2)^2 = 35/12.$$

The required statistics are therefore

$\langle f(n) \rangle = \frac{7}{2}$,
$\langle f(n_1) f(n_2) \rangle = \frac{49}{4} + \frac{35}{12} \delta_{n_1, n_2}$,
$C(n_1, n_2) = \frac{35}{12} \delta_{n_1, n_2}.$

Solution 2.2

No solution is given, because there are many possible outcomes.

Solution 2.3

For diffusive motion, the typical displacement is proportional to the square root of the time. The number of seconds in one week is $60 \times 60 \times 24 \times 7 = 604\,800$. For the random walk, the typical displacement is therefore $\sqrt{604\,800}\,\text{mm} \simeq 778\,\text{mm} = 0.778\,\text{m}$.

For constant velocity $1\,\text{mm}\,\text{s}^{-1}$, the distance travelled in one week is $604\,800\,\text{mm} \simeq 605\,\text{m}$.

Solution 2.4

Following the same approach as in Example 2.3, we define a random function $f(n)$ which takes the value $+1$ with probability p, or -1 with probability $1-p$. The statistics of this random function were given in Example 2.1:

$$\langle f(n) \rangle = 2p - 1$$

and

$$\langle f(n_1) f(n_2) \rangle = \begin{cases} (2p-1)^2, & \text{for } n_1 \neq n_2, \\ 1, & \text{for } n_1 = n_2. \end{cases}$$

Calculating the statistics of $X(T)$ using the approach of Example 2.3 gives

$$\langle X(T) \rangle = (2p-1)T,$$

$$\langle [X(T)]^2 \rangle = (T^2 - T)(2p-1)^2 + T,$$

$$\begin{aligned}\text{Var}(X) &= (T^2 - T)(2p-1)^2 + T - T^2(2p-1)^2 \\ &= [1 - (2p-1)^2]T \\ &= 4p(1-p)T.\end{aligned}$$

Again, the variance is proportional to T, so this process is a random walk, now with diffusion constant $D = 2p(1-p)$. Also, the mean displacement is proportional to T, and the drift velocity is $v = 2p - 1$. These results agree with those for the simple random walk when $p = \frac{1}{2}$.

Solution 2.5

The non-zero entries in the row for $T = 5$ are

$$P(-5, 5) = \tfrac{1}{2}P(-4, 4) = \tfrac{1}{32},$$
$$P(-3, 5) = \tfrac{1}{2}[P(-4, 4) + P(-2, 4)] = \tfrac{1}{2}(\tfrac{1}{16} + \tfrac{4}{16}) = \tfrac{5}{32},$$
$$P(-1, 5) = \tfrac{1}{2}[P(-2, 4) + P(0, 4)] = \tfrac{10}{32},$$

and by symmetry $P(1, 5) = \tfrac{10}{32}$, $P(3, 5) = \tfrac{5}{32}$, $P(5, 5) = \tfrac{1}{32}$.

For $T = 6$: $P(-6, 6) = \tfrac{1}{64}$, $P(-4, 6) = \tfrac{6}{64}$, $P(-2, 6) = \tfrac{15}{64}$, $P(0, 6) = \tfrac{20}{64}$, and for positive X we can again use the symmetry $P(X, T) = P(-X, T)$.

As a further check, the entries in each row should add up to one.

Solution 2.6

The numbers in the next row are 1, 6, 15, 20, 15, 6, 1.

Also, $(1 + x)^6 = 1 + 6x + 15x^2 + 20x^3 + 15x^4 + 6x^5 + x^6$, so the coefficients of x^r are the elements of the seventh row of Pascal's triangle, as expected.

Solution 2.7

We demonstrate that the binomial coefficients satisfy the recurrence relation:

$$\begin{aligned}
C_r^{n-1} + C_{r-1}^{n-1} &= \frac{(n-1)!}{r!\,(n-1-r)!} + \frac{(n-1)!}{(r-1)!\,(n-r)!} \\
&= \frac{(n-1)!}{r!\,(n-r)!}[(n-r) + r] \\
&= \frac{n!}{r!\,(n-r)!} = C_r^n.
\end{aligned}$$

Now we apply this result to the proposed solution of the recurrence relation for $P(X, T)$:

$$\begin{aligned}
&\tfrac{1}{2}\left[P(X-1, T-1) + P(X+1, T-1)\right] \\
&= \frac{1}{2}\left[\frac{1}{2^{T-1}} C_{(X-1+T-1)/2}^{T-1} + \frac{1}{2^{T-1}} C_{(X+1+T-1)/2}^{T-1}\right] \\
&= \frac{1}{2}\left[\frac{1}{2^{T-1}} C_{(X+T)/2-1}^{T-1} + \frac{1}{2^{T-1}} C_{(X+T)/2}^{T-1}\right] \\
&= \frac{1}{2^T}\left[C_{(X+T)/2-1}^{T-1} + C_{(X+T)/2}^{T-1}\right] \\
&= \frac{1}{2^T} C_{(X+T)/2}^T \\
&= P(X, T),
\end{aligned}$$

where the penultimate equality follows directly from the first part of the question.

Solution 2.8

Following the same reasoning as was used to derive equation (2.27), the equation is

$$P(X, T) = p\,P(X-1, T-1) + (1-p)\,P(X+1, T-1).$$

Solution 2.9

Using the approximate distribution (2.32), the second moment of X is

$$\langle X^2 \rangle \simeq \sum_X P_{\text{app}}(X,T)\, X^2 = \frac{2}{\sqrt{2\pi T}} \sum_X X^2 \exp(-X^2/2T),$$

where the sum is over integer values of X with the same parity (odd or even) as T. A summation may be approximated by an integral, as

$$\sum_n \delta x\, f(n\,\delta x) \simeq \int_a^b dx\, f(x),$$

where the sum is over all values of n such that $a < n\,\delta x < b$. This is a good approximation when δx is small relative to the scale over which the function $f(x)$ changes.

When T is large, the function $P_{\text{app}}(X,T)$ varies very slowly as a function of X. We set $\delta x = 2$ in the above expression, because adjacent values of X differ by two. The summation is then approximated by an integral, divided by $\delta x = 2$:

$$\langle X^2 \rangle \simeq \frac{2}{\sqrt{2\pi T}} \frac{1}{2} \int_{-\infty}^{\infty} dX\, X^2 \exp(-X^2/2T)$$
$$= \frac{1}{\sqrt{2\pi T}} (2T)^{3/2} \int_{-\infty}^{\infty} du\, u^2 \exp(-u^2) = \frac{2T}{\sqrt{\pi}} I_2 = T,$$

The integral I_2 is discussed in Section 1.3.

where the substitution $u = X/\sqrt{2T}$ was used to get from the first integral to the second. The final result is identical to the exact one given by equation (2.17).

Solution 2.10

By Stirling's formula, $\ln(6!) = 6.5653\ldots$, so $6! \simeq 710$. The correct value is $6! = 720$. The relative error is $10/720 \simeq 0.014$.

Similarly, Stirling's formula gives $\ln(8!) = 10.5941\ldots$ and $\ln(10!) = 15.0960\ldots$, giving the approximate values $8! \simeq 39\,902$ and $10! \simeq 3.5987 \times 10^6$. The exact values are $8! = 40\,320$ and $10! = 3\,628\,800$, so the relative errors are approximately 0.010 and 0.008, respectively.

The relative error is seen to decrease as N increases.

Solution 2.11

By inspection,

$$\frac{d}{dx}[x\ln(x) - x] = \ln(x),$$

so the indefinite integral of $\ln(x)$ is $x\ln(x) - x$.

Solution 2.12

After applying Stirling's formula, and ignoring the error terms, the first line of equation (2.39) becomes

$$\ln[P(X,T)] \simeq T\ln(T) - T + \tfrac{1}{2}\ln(2\pi T) - T\ln(2)$$
$$- \left(\frac{T+X}{2}\right)\ln\left(\frac{T+X}{2}\right) + \left(\frac{T+X}{2}\right) - \tfrac{1}{2}\ln[\pi(T+X)]$$
$$- \left(\frac{T-X}{2}\right)\ln\left(\frac{T-X}{2}\right) + \left(\frac{T-X}{2}\right) - \tfrac{1}{2}\ln[\pi(T-X)],$$

which can be rearranged to give the final line.

Solutions to Exercises in Chapter 2

Solution 2.13

Using the trial solution $P_{\text{app}}(X,T) = 2\exp(-X^2/2T)/\sqrt{2\pi T}$, we have

$$\frac{\partial P_{\text{app}}}{\partial X}(X,T) = -\frac{X}{T}\frac{2}{\sqrt{2\pi T}}\exp(-X^2/2T) = -\frac{X}{T}P_{\text{app}}(X,T),$$

$$\frac{\partial^2 P_{\text{app}}}{\partial X^2}(X,T) = \left[\frac{X^2}{T^2} - \frac{1}{T}\right]P_{\text{app}}(X,T),$$

$$\frac{\partial P_{\text{app}}}{\partial T}(X,T) = \left[\frac{X^2}{2T^2} - \frac{1}{2T}\right]\frac{2}{\sqrt{2\pi T}}\exp(-X^2/2T) = \frac{1}{2}\left[\frac{X^2}{T^2} - \frac{1}{T}\right]P_{\text{app}}(X,T).$$

It follows that this trial solution satisfies

$$\frac{\partial P_{\text{app}}}{\partial T} = \frac{1}{2}\frac{\partial^2 P_{\text{app}}}{\partial X^2},$$

which is the diffusion equation with $D = \frac{1}{2}$.

Solution 2.14

The function $f(n)$ has the statistics $\langle f(n)\rangle = 0$ and $\langle f(n_1)f(n_2)\rangle = \delta_{n_1,n_2}$. The mean value of the function $g(n)$ is

$$\langle g(n)\rangle = \left\langle \sum_{m=-\infty}^{\infty} w(n-m)f(m) \right\rangle = \sum_{m=-\infty}^{\infty} w(n-m)\langle f(m)\rangle = 0.$$

(Note that the $w(m)$ are fixed numbers, not random variables, so $\langle w(n-m)f(m)\rangle = w(n-m)\langle f(m)\rangle$.) The correlation function $C(n_1, n_2)$ of $g(n)$ is

$$\langle g(n_1)g(n_2)\rangle = \left\langle \sum_{m_1=-\infty}^{\infty}\sum_{m_2=-\infty}^{\infty} w(n_1-m_1)w(n_2-m_2)f(m_1)f(m_2) \right\rangle$$

$$= \sum_{m_1=-\infty}^{\infty}\sum_{m_2=-\infty}^{\infty} w(n_1-m_1)w(n_2-m_2)\langle f(m_1)f(m_2)\rangle$$

$$= \sum_{m_1=-\infty}^{\infty}\sum_{m_2=-\infty}^{\infty} w(n_1-m_1)w(n_2-m_2)\delta_{m_1,m_2}$$

$$= \sum_{m_1=-\infty}^{\infty} w(n_1-m_1)w(n_2-m_1)$$

$$= \sum_{m=-\infty}^{\infty} w(n_1-n_2+m)w(m).$$

In the last line, the summation variable was changed to $m = n_2 - m_1$.

Note that this is a function of the difference $n_1 - n_2$, rather than of n_1 and n_2 separately, so the correlation function depends upon only one variable and can be written as $C(n_2 - n_1)$.

Solution 2.15

The sum of the values of the correlation function from the previous exercise is

$$S = \sum_{n=-\infty}^{\infty} C(n)$$

$$= \sum_{n=-\infty}^{\infty}\sum_{m=-\infty}^{\infty} w(m-n)w(m)$$

$$= \sum_{m=-\infty}^{\infty} w(m) \sum_{k=-\infty}^{\infty} w(k)$$

$$= \left[\sum_{n=-\infty}^{\infty} w(n)\right]^2 \geq 0,$$

In the penultimate line, the substitution $m - n = k$ was used.

where equality to zero holds only if $w(n) = 0$ for all n.

Solution 2.16

Assume that $T_1 \leq T_2$. Then

$$\langle X(T_1)X(T_2)\rangle = \left\langle \sum_{n_1=1}^{T_1} f(n_1) \sum_{n_2=1}^{T_2} f(n_2) \right\rangle$$

$$= \sum_{n_1=1}^{T_1} \sum_{n_2=1}^{T_2} \langle f(n_1)f(n_2)\rangle$$

$$= \sum_{n_1=1}^{T_1} \sum_{n_2=1}^{T_2} \delta_{n_1,n_2} = \sum_{n_1=1}^{T_1} 1 = T_1.$$

Swapping the names of the symbols T_1 and T_2, we conclude that if $T_2 \leq T_1$ then this statistic is equal to T_2. It follows that $\langle X(T_1)X(T_2)\rangle = \min(T_1, T_2)$, as required. Note that the summation over n_2 is equal to unity only if n_1 is contained in the set of integers $\{1, 2, \ldots, T_2\}$, so if we had not assumed that $T_1 \leq T_2$, the final line would have been incorrect.

Alternatively, consider Figure 2.12, where $T_2 < T_1$, which shows that there are only T_2 non-zero values in the double sum.

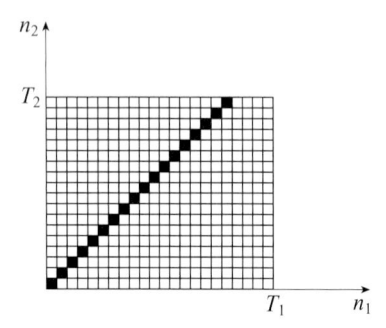

Figure 2.12 Diagram showing values of (n_1, n_2) contributing to the double sum in Exercise 2.16 when $T_2 < T_1$. The T_2 non-zero values of $\langle f(n_1)f(n_2)\rangle$ are shown as black squares.

Solution 2.17

When the numbers n_1, n_2, n_3, n_4 are all different, the random numbers $f(n_1)$, $f(n_2)$, $f(n_3)$ and $f(n_4)$ are independent, and the mean value $\langle f(n_1)f(n_2)f(n_3)f(n_4)\rangle$ can be written as a product of four factors, each of which is zero. Similarly, if only two or three of these numbers are the same, the mean value with respect to the remaining number or numbers is equal to zero. The only possibility for the mean value being non-zero is if the numbers are equal in pairs: the three possible pairings are $n_1 = n_2$, $n_3 = n_4$ or $n_1 = n_3$, $n_2 = n_4$ or $n_1 = n_4$, $n_2 = n_3$. Considering one of these cases ($n_1 = n_3$, $n_2 = n_4$, say), we obtain

$$\langle [f(n_1)]^2 [f(n_2)]^2\rangle = \langle 1 \times 1\rangle = 1.$$

In the case where $n_1 = n_2 = n_3 = n_4 = n$, we have $\langle [f(n)]^4\rangle = 1$, because $[f(n)]^4$ is always unity. Thus the mean value $\langle f(n_1)f(n_2)f(n_3)f(n_4)\rangle$ is equal to unity if the arguments are equal in pairs, and is zero otherwise. This conclusion may be expressed by writing

$$\langle f(n_1)f(n_2)f(n_3)f(n_4)\rangle = \delta_{n_1,n_2}\delta_{n_3,n_4} + \delta_{n_1,n_3}\delta_{n_2,n_4} + \delta_{n_1,n_4}\delta_{n_2,n_3}$$
$$- 2\delta_{n_1,n_2}\delta_{n_1,n_3}\delta_{n_1,n_4}.$$

The first three of the right-hand-side factors are unity when the arguments are paired, and zero otherwise. The final term is non-zero only when all arguments are equal, and in this case $\langle [f(n)]^4\rangle = 1 + 1 + 1 - 2 = 1$, as required.

Given this result, we have

$$\langle [X(T)]^4 \rangle = \sum_{n_1=1}^{T} \sum_{n_2=1}^{T} \sum_{n_3=1}^{T} \sum_{n_4=1}^{T} \langle f(n_1)f(n_2)f(n_3)f(n_4) \rangle$$

$$= \sum_{n_1=1}^{T} \sum_{n_2=1}^{T} \sum_{n_3=1}^{T} \sum_{n_4=1}^{T} (\delta_{n_1,n_2}\delta_{n_3,n_4} + \delta_{n_1,n_3}\delta_{n_2,n_4} + \delta_{n_1,n_4}\delta_{n_2,n_3}$$
$$- 2\delta_{n_1,n_2}\delta_{n_1,n_3}\delta_{n_1,n_4})$$

$$= \sum_{n_1=1}^{T} 1 \sum_{n_3=1}^{T} 1 + \sum_{n_1=1}^{T} 1 \sum_{n_2=1}^{T} 1 + \sum_{n_1=1}^{T} 1 \sum_{n_2=1}^{T} 1 - 2\sum_{n_1=1}^{T} 1$$

$$= T^2 + T^2 + T^2 - 2T$$

$$= 3T^2 - 2T.$$

When studying Gaussian distributions, we found that $\langle x^4 \rangle = 3[\langle x^2 \rangle]^2$. For the simple random walk, we have $\langle X^2 \rangle = T$, so if the distribution has a Gaussian probability density, we would expect $\langle X^4 \rangle = 3T^2$. This is equivalent to the exact result at leading order in T.

See Exercise 1.28.

Solution 2.18

Setting $\epsilon = X/T$, we have

$$\left(\frac{T+X}{2}\right)\ln\left(\frac{T+X}{2}\right) = \frac{T}{2}(1+\epsilon)\ln\left(\frac{T}{2}(1+\epsilon)\right)$$

$$= \frac{T}{2}(1+\epsilon)\left[\ln\left(\frac{T}{2}\right) + \ln(1+\epsilon)\right]$$

$$= \frac{T}{2}\ln\left(\frac{T}{2}\right) + \frac{T}{2}\ln(1+\epsilon) + \left[\frac{T}{2}\ln\left(\frac{T}{2}\right)\right]\epsilon + \frac{T\epsilon}{2}\ln(1+\epsilon)$$

$$= \frac{T}{2}\ln\left(\frac{T}{2}\right) + \frac{T}{2}\left[\epsilon - \frac{\epsilon^2}{2} + \frac{\epsilon^3}{3} + O(\epsilon^4)\right] + \left[\frac{T}{2}\ln\left(\frac{T}{2}\right)\right]\epsilon$$
$$+ \frac{T\epsilon}{2}\left[\epsilon - \frac{\epsilon^2}{2} + \frac{\epsilon^3}{3} + O(\epsilon^4)\right]$$

$$= \frac{T}{2}\ln\left(\frac{T}{2}\right) + \frac{T}{2}\left[1 + \ln\left(\frac{T}{2}\right)\right]\epsilon + \frac{T}{4}\epsilon^2 + O(\epsilon^3).$$

Similarly, by changing the sign of ϵ,

$$\left(\frac{T-X}{2}\right)\ln\left(\frac{T-X}{2}\right) = \frac{T}{2}(1-\epsilon)\ln\left(\frac{T}{2}(1-\epsilon)\right)$$

$$= \frac{T}{2}\ln\left(\frac{T}{2}\right) - \frac{T}{2}\left[1 + \ln\left(\frac{T}{2}\right)\right]\epsilon + \frac{T}{4}\epsilon^2 + O(\epsilon^3).$$

Substituting these results into equation (2.39) gives

$$\ln[P(X,T)] \simeq T\ln(T) - \left(\frac{T}{2}\ln\left(\frac{T}{2}\right) + \frac{T}{2}\left[1 + \ln\left(\frac{T}{2}\right)\right]\epsilon + \frac{T}{4}\epsilon^2\right)$$
$$- \left(\frac{T}{2}\ln\left(\frac{T}{2}\right) - \frac{T}{2}\left[1 + \ln\left(\frac{T}{2}\right)\right]\epsilon + \frac{T}{4}\epsilon^2\right) - T\ln(2)$$
$$+ \tfrac{1}{2}\ln\left[\frac{2}{\pi T(1+\epsilon)(1-\epsilon)}\right]$$

$$\simeq T\ln(T) - T\ln\left(\frac{T}{2}\right) - T\ln(2) - \frac{T}{2}\epsilon^2 + \tfrac{1}{2}\ln\left[\frac{2}{\pi T(1-\epsilon^2)}\right].$$

We therefore have

$$\ln[P(X,T)] \simeq -\frac{T}{2}\epsilon^2 + \tfrac{1}{2}\ln\left(\frac{2}{\pi T}\right) - \tfrac{1}{2}\ln(1-\epsilon^2).$$

Now using equation (2.40), $\ln(1-\epsilon^2)$ is approximated by $-\epsilon^2$, giving

$$\ln[P(X,T)] \simeq \tfrac{1}{2}\ln\left(\frac{2}{\pi T}\right) - \left(\frac{T-1}{2}\right)\epsilon^2.$$

When $T \gg 1$,
$$\ln[P(X,T)] \simeq \tfrac{1}{2}\ln\left(\frac{2}{\pi T}\right) - \frac{T}{2}\epsilon^2,$$
which is equation (2.41).

CHAPTER 3
The diffusion equation

3.1 Introduction

In this chapter we shall explain a derivation of the diffusion equation, as a description both of the process of diffusion of dissolved materials, and of changes in temperature due to the flow of heat. The derivation is based upon a plausible assumption, which is known as Fick's law (for particle diffusion) or Fourier's law (for the flow of heat). A more fundamental derivation, based upon a random walk model for the motion of molecules, will be considered in Chapter 6. Before introducing the physical concepts which lead to the diffusion equation, we shall discuss the mathematical form of this equation.

The diffusion equation is a linear partial differential equation. The one-dimensional form of the diffusion equation has already been introduced in Chapter 2, where it appears as equation (2.43). It is reproduced below with a slightly different notation:

$$\frac{\partial f}{\partial t} = D\frac{\partial^2 f}{\partial x^2}. \tag{3.1}$$

Here D is a positive constant known as the *diffusion constant*. A function $f(x,t)$ must be determined which satisfies this equation. As it stands, the equation does not have a unique solution. Additional information, in the form of conditions which must be satisfied by the function $f(x,t)$, is determined by the system that the equation is used to model. These additional conditions are usually termed *initial data* and *boundary conditions*, and will be discussed in detail in Chapter 4.

The function $f(x,t)$ and the variables x and t may represent various quantities in different contexts. When the equation is used to describe diffusion, f represents a probability density or a *concentration* of particles (which will be described shortly). The variable x usually represents a Cartesian coordinate for the position of a point in space, but diffusion can occur in more abstract spaces; for example, in applications to financial mathematics, x could represent the value of a commodity. The variable t almost always represents time. When the equation is used to describe the flow of heat, f represents the temperature at position x and time t. In this context the diffusion equation is known as the *heat equation*.

In three dimensions, the diffusion equation takes the form

$$\frac{\partial f}{\partial t} = D\nabla^2 f, \tag{3.2}$$

where f depends upon position $\boldsymbol{r} = (x, y, z)$ and time t, and where ∇^2 is the *Laplacian operator*. Its action on a function $f(x, y, z)$ is defined by

$$\nabla^2 f = \frac{\partial^2 f}{\partial x^2} + \frac{\partial^2 f}{\partial y^2} + \frac{\partial^2 f}{\partial z^2}. \tag{3.3}$$

∇^2 is pronounced 'nabla squared'.

In some texts, the Laplacian of f is written Δf, rather than $\nabla^2 f$. The symbol Δ will be used for another purpose in this chapter. The Laplacian operator was introduced earlier, in Chapter 2 of Block 0.

Δ is the upper case Greek 'delta'.

Exercise 3.1

Calculate $\nabla^2 f$ for each of the following functions.

(a) $f(x, y, z) = x^2 + y^2 + z^2$

(b) $f(x, y, z) = x^2 - y^2$

(c) $f(\boldsymbol{r}) = 1/r$, where $r = |\boldsymbol{r}| = \sqrt{x^2 + y^2 + z^2}$.
(Do not consider the point where $r = 0$, where f is not differentiable.)

Part (c) requires a fair amount of algebra to arrive at a very simple result. A more insightful route to the same conclusion will be considered later, in Exercise 3.17.

Exercise 3.2

(a) Based upon Equations (3.1) and (3.2), guess the form of the diffusion equation in two dimensions, with coordinates (x, y) and diffusion constant D.

(b) Find a solution of the two-dimensional diffusion equation in the form $\rho(x, y, t) = A(t) \exp[-\beta(x^2 + y^2)/t]$, where the constant β is to be determined. What is the form of the function $A(t)$?

[Hint: Substitute this expression into the diffusion equation, and group terms which have the same dependence on position. You will find a relation between β and D, and a differential equation for $A(t)$.]

This part of the exercise will develop your facility with partial differentiation and simple ordinary differential equations. The solution to Exercise 3.24 will discuss a more direct alternative approach.

Having described the mathematical form of the diffusion equation, most of the remainder of the chapter will consider how it arises in a physical context. This requires introducing a sequence of concepts: concentration, flux, flux density, and the continuity equation. Having discussed these concepts, we then introduce Fick's law, and use it to derive the diffusion equation.

3.2 Diffusion

Scientific enquiry has demonstrated that all materials are made up of very small particles called atoms or molecules. (Molecules are clusters of atoms tightly bound together by 'chemical bonds', but the distinction between atoms and molecules is not important for understanding the mathematical content of what follows.) These particles are so small that they are invisible, even using a microscope. Any piece of material that you can see with the naked eye therefore contains enormous numbers of particles; for example, a glass of water contains (very roughly) 10^{23} water molecules.

Even in a material where there is no detectable motion to the naked eye, the molecules are always in motion. The motion of the molecules in liquids and gasses (illustrated in the previous chapter in Figure 2.4) is apparently

3.2 Diffusion

random, in all possible directions, and with varying speeds (typically hundreds of metres per second, for materials at room temperature). Atoms in solids are also constantly in motion, but they vibrate about fixed positions.

Our everyday experience is that objects do not remain forever in motion, but lose energy due to the effects of friction. When two atoms collide, however, it is now accepted that the total kinetic energy is 'conserved', meaning that the total kinetic energy after the collision is the same as before. Because no energy is lost when atoms collide, the collisions do not result in the atoms slowing down. It is natural to ask why atoms are unaffected by friction, whereas *macroscopic* (larger scale) objects are subject to friction. The explanation follows from the fact that energy which is apparently lost into friction is the result of the kinetic energy of a large body being converted into the apparently random motions of many atoms. In the case of atoms, there are no smaller constituents into which their kinetic energy can be distributed, so that collisions between atoms involve no loss of energy.

The rapid motion of atoms or molecules is described as happening on a *microscopic* scale, even though it cannot be observed directly even with the most powerful optical microscope. (The distance between the molecules in a liquid might be 10^{-9} m or less, whereas a microscope using visible light does not allow us to distinguish objects much less than 10^{-6} m across.)

Although the microscopic motion of atoms cannot be seen directly, it gives rise to phenomena that are observable on a larger ('macroscopic') scale. Diffusion is one such phenomenon: it causes mixing of different types of atoms or molecules. Figure 3.1 illustrates a process of diffusion. One layer of liquid (clear) is placed on top of another layer (dark). After a while, the boundary between these layers is no longer distinct. The explanation is that some of the molecules in the coloured liquid have migrated up into the clear layer, and conversely some molecules from the clear layer have moved downwards. This mixing occurs as a result of the microscopic motion of the molecules.

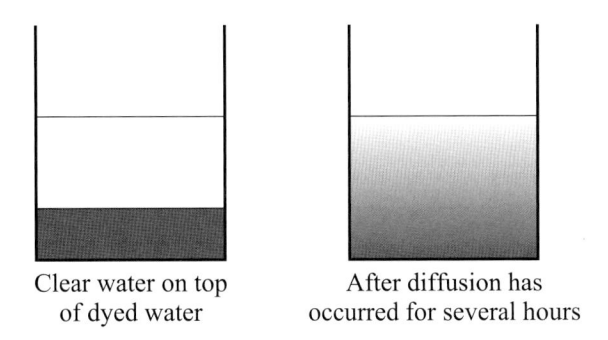

Figure 3.1 Diffusion of one liquid into another

> Mixing of fluids can involve processes other than diffusion. Mixing by diffusion occurs when there is no detectable flow of the fluid. If the fluids are stirred, they flow in a complex way that greatly enhances the mixing.
>
> Another way of mixing fluids is by *thermal convection* in which temperature differences create differences in density, and the less dense fluid flows to the surface. In most fluids, density decreases as temperature increases, so if a fluid is heated from below, the warm fluid rises away from the source of heat. This causes a mixing process, called thermal convection. Where flow of the fluid is present, this normally mixes substances far more effectively than diffusion alone.
>
> If you want to try the experiment illustrated in Figure 3.1 yourself, you must make sure that the dyed and clear water do not mix by thermal convection:

keep the beaker on a cold surface, such as a stone floor, in order to avoid this. A method for observing diffusion by a simple experiment that you can try at home is discussed in the Appendix to this chapter.

Diffusion occurs in gases as well as liquids. Sometimes it is claimed that when a bottle of perfume is opened on one side of a room, the smell reaches the other side by a process of diffusion. This is possible in principle, but in practice it applies only in rooms where the air is unusually still. The scent is usually carried by imperceptible air currents, set up by thermal convection, which distribute the scent more rapidly than by diffusion alone.

Trying to calculate the motion of individual molecules is usually an impractical approach, and statistical methods must be used to understand the diffusion process fully. The statistical description of microscopic motion will be considered in Chapter 6. This chapter considers only a macroscopic description of diffusion, based upon a natural assumption about the diffusion process which is called *Fick's law*. In order to discuss this law, we must first introduce the concepts of concentration and flux density.

3.3 Concentration and flux density

When describing diffusion we need some measure of the quantity of a substance which is present at some point at a given time. This is the *concentration*, denoted by c, which is a function of position r and time t. We also need a measure of the flow of material. This is the *flux density* J; it is a vector quantity, because the flow of material has a direction, and is also a function of r and t.

The concentration and the flux density are most easily defined for situations where the system is *homogeneous*, meaning that the distribution of particles is expected to be equivalent at all points in space, so that c and J are independent of the position r. In each case that we consider, the definitions of both c and J will be given for a homogeneous system in the first instance, before considering the more difficult general situation.

3.3.1 Concentration

Let us consider first the concentration, c. In the case where a fluid is mixed (for example, by stirring it) so that it becomes homogeneous, the concentration of a particular type of atom (or molecule, or other particle) in the fluid is simply the ratio of the number of atoms N to the volume of fluid V, so

$$c = \frac{N}{V}. \tag{3.4}$$

In the case where the system is not homogeneous, the concentration $c(r, t)$ at position r and time t may be defined as follows. Consider a small element of volume ΔV, with its centre at r. The number of molecules of a particular type in this region is $\Delta N(r, t)$ at time t. The concentration is defined as

$$c(r,t) = \frac{\Delta N(r,t)}{\Delta V}. \tag{3.5}$$

The concentration as defined by equation (3.5) depends upon the choice of ΔV. Clearly ΔV should be small if $c(r, t)$ is to faithfully represent what is happening at position r (see Figure 3.2). It is tempting to define $c(r, t)$

3.3 Concentration and flux density

in terms of the limit of $\Delta N/\Delta V$ as $\Delta V \to 0$, but that limit would be zero for most points in space, because almost all points do not coincide with the position of a particle. In practice, it is sufficient to choose an extremely small volume ΔV, and the number of particles contained in this volume will still be very large: the numbers of atoms in even the tiniest speck of material visible to the naked eye are extremely large. The value of the concentration will be, for all practical purposes, independent of the choice of ΔV, provided that ΔV is neither too large nor too small.

> The shape of the volume element is not important, provided that it is not highly elongated in any direction. It could be taken to be a sphere or a cube, for example.
>
> Throughout this chapter, the symbols δ and Δ (lower and upper case Greek 'delta') will be used to indicate small amounts of some quantity. Thus V stands for volume and ΔV indicates a small volume, and δt indicates a small time. Sometimes, there will be two distinct small quantities considered, so that (for example) δN and ΔN may represent two different small numbers of particles in the same calculation.
>
> When we discuss numbers of particles, the quantity ΔN is small only in the sense that it is much smaller than the total number of particles N. It will become clear shortly that the volume element ΔV should be sufficiently large that $\Delta N \gg 1$.

Exercise 3.3

(This exercise is to check your understanding of the concept of concentration: no calculation is required.)

What is the appropriate definition of concentration in one and two dimensions?

Atlases often contain data on population densities; for example, in 1983 Hong Kong had a population density of 5300 per square kilometre, and the United Kingdom had a population density of 230 per square kilometre. How is this concept of population density related to concentration?

The definition of concentration is illustrated (in the two-dimensional case) in Figure 3.2.

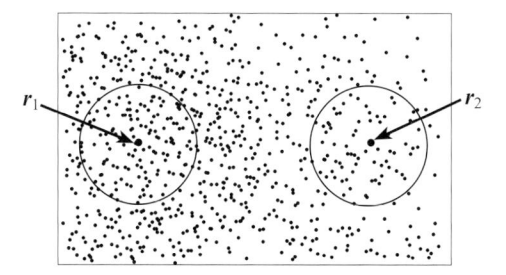

Figure 3.2 Illustrating the definition of concentration. In this two-dimensional example, the concentration at a point r is defined as the number of particles inside a disk centred at r, divided by the area of the disk. The concentration of particles at r_1 is higher than that at r_2.

It is useful to be able to express the total number of particles inside a volume V in terms of the concentration. If the system is homogeneous, equation (3.4) implies that the number of particles is $N = cV$. Let us consider how this expression must be modified when c is not uniform across the volume. We can divide the volume up into small elements, labelled by $i = 1, \ldots, M$, each element i having volume ΔV_i. The number of particles

ΔN_i within element i is approximately $c_i \Delta V_i$, where c_i is the concentration at the centre of that element. The total number of particles is then

$$N = \sum_{i=1}^{M} \Delta N_i \simeq \sum_{i=1}^{M} c_i \Delta V_i. \tag{3.6}$$

If the concentration is a known function of position, we can make the volume elements shrink in size, while still covering the volume V. The integral over a volume is defined in terms of such a limit, so the number of particles inside V at time t is given by the volume integral (as defined in Chapter 2 of Block 0):

$$N = \int_V dV \ c(\boldsymbol{r}, t). \tag{3.7}$$

Exercise 3.4

Suppose that the concentration of particles in a one-dimensional region between $x = 0$ and $x = L$ is

$c(x, t) = c_0 + c_1 \cos(\pi x/L) \exp(-\alpha t),$

where c_0, c_1 and α are constants. Calculate the total number of particles $N(a)$ between $x = 0$ and $x = a$ (where $a \leq L$). Verify that $N(a)$ is independent of time when $a = L$.

3.3.2 Molar concentration

Because the numbers of atoms in macroscopic objects are extremely large, it is convenient to quote numbers of atoms or molecules relative to a large number, the *mol* (pronounced 'mole').

In most of the exercises, quantities will be represented by symbols which will not be converted into numbers, but in many practical applications scientists use *molar* concentrations, so it is important to be familiar with the concept.

One mol is (approximately) equal to $6.022\,045 \times 10^{23}$. Thus, for example, $0.034 \,\mathrm{mol} \simeq 2.047 \times 10^{22}$.

Exercise 3.5

If 10^{20} molecules of methanol are disolved in 500 ml of water and stirred until the solution becomes homogeneous, what is the concentration of methanol in the water? Express the answer as the number of molecules per cubic metre, and as the number of mols per litre. [Hint: $1\,\mathrm{m}^3 = 10^3\,\mathrm{l}$.]

The number used in the definition of the mol is known as Avogadro's number, $N_A \simeq 6.022 \times 10^{23}$. You might well wonder why such an awkward number is used. The definition of the mol is related to the mass of a carbon-12 atom. The mol is defined by the statement that 1 mol of carbon-12 atoms has a mass of 12 g. You need not remember the definition of the mol, but should be aware of its existence. The weights of atoms and molecules relative to carbon-12 can be found from chemical tables, and the molar quantity of a substance is easily obtained from its weight.

The original intention had been to define the mol as the number of atoms in one gram of hydrogen, but for technical reasons it is easier to count carbon atoms.

Like other chemical elements, carbon atoms occur as different *isotopes*, having different mass: most carbon atoms in nature are of the carbon-12 type.

Exercise 3.6

The concentration of a substance can also be expressed as a mass per unit volume, that is, as a density. The air in a room might contain carbon dioxide at a concentration of $250\,\mu\mathrm{g\,l}^{-1}$ (250 micrograms per litre), and carbon monoxide at $2\,\mu\mathrm{g\,l}^{-1}$. Each carbon dioxide molecule has mass 7.31×10^{-26} kg and each carbon monoxide molecule has mass 4.65×10^{-26} kg. Express these concentrations in terms of numbers of particles per cubic metre.

To summarise: in practical applications, a concentration may be quoted as the number of particles per unit volume, as mols per unit volume, or as mass per unit volume.

3.3.3 Flux and flux density

Having defined the concentration, our objective will be to describe its variation in time and space by means of a partial differential equation, the diffusion equation. This equation will be derived by considering the rate at which particles enter or leave a small volume element. This rate is described using a quantity known as the *flux density*: this will be defined as a scalar quantity J, then this will be generalised so that it is described by a vector $\mathbf{J}$.

First, a flux Φ of particles will be defined, associated with a given surface, and this will then be used to define the flux density. The flux is the number of particles passing through the surface per unit time. (The surface need not be a physical barrier, just a mathematical construct, having no influence on the motion of the particles.) If n particles pass through the surface in time t, the flux is

$$\Phi = \frac{n}{t}. \tag{3.8}$$

This formula is appropriate when the flow is *steady*, that is, if the flow does not vary with time. Later, we shall give expressions that are valid when the flow is changing, so that Φ is a function of time.

The definition of flux requires us to distinguish the two sides of the surface: if the two sides are labelled A and B, say, we consider n to be the net number of particles passing from side A to side B, that is, the number passing from A to B, minus the number passing from B to A.

The flux density quantifies how rapidly particles cross a given surface, in terms of the rate of flow per unit area. In the case of a homogeneous fluid moving across a flat surface of area A at uniform velocity, the flux density is independent of position and time, and is defined as

$$J = \frac{\Phi}{A}. \tag{3.9}$$

The definitions of flux and flux density for a steady homogeneous flow are illustrated in Figure 3.3.

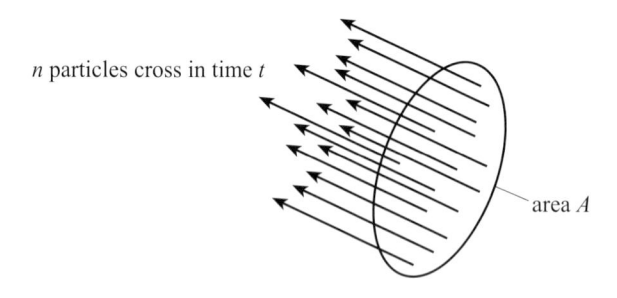

Figure 3.3 Illustrating the definition of flux and flux density for a steady homogeneous flow. If n particles pass through this surface in time t, the flux is $\Phi = n/t$. If the area of the surface is A, the flux density is $J = \Phi/A$.

Exercise 3.7

A net is placed across a river, so that it catches all of the fish swimming along it. If 600 fish are caught in one day, what is the flux of fish? If the river is 5 m wide, and the water is 1 m deep, what is the flux density of fish? (Assume a steady homogeneous flow.)

Exercise 3.8

During a hailstorm lasting 5 minutes, 100 g of hailstones are collected from a circular pan with a diameter of 20 cm. Some hailstones are weighed, and their average weight is found to be 0.05 g. Estimate the average flux density of hailstones across any horizontal surface during this storm. (Calculate your answer in the appropriate SI unit.)

In general, the flux density is defined to be a function of position r and time t. At position r, we place a small flat or nearly flat surface element of area ΔA. The normal to this surface is a vector n of unit length which is perpendicular to this surface. The surface has two sides, A and B say, and we distinguish these by saying that the vector n points from side A towards side B. Between time t and time $t + \delta t$, the particles crossing this surface are observed and counted. The number of particles crossing in the direction from A to B minus the number crossing in the opposite direction is δn (see Figure 3.4). The flux of particles through this small surface element is $\Delta \Phi = \delta n / \delta t$.

The scalar flux density J of particles in the direction of n is the ratio of the small flux $\Delta \Phi$ to the small area ΔA:

$$J(\boldsymbol{n}, \boldsymbol{r}, t) = \frac{\Delta \Phi}{\Delta A} = \frac{\delta n}{\Delta A \, \delta t}. \tag{3.10}$$

This quantity depends upon the choice of ΔA and δt. However, when the number of particles involved is very large, for a wide range of small values of ΔA and δt the value of $J(\boldsymbol{n}, \boldsymbol{r}, t)$ will be almost independent of the values of these quantities. It is also possible to conceive of very irregular flows where J is not well-defined, but we shall not consider these cases.

3.3 Concentration and flux density

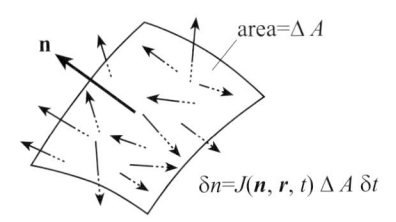

Figure 3.4 Illustrating the definition of flux density: δn is the net number of particles passing through the surface element in time δt

Now we consider how the vector flux density $\boldsymbol{J} = J_1\mathbf{i} + J_2\mathbf{j} + J_3\mathbf{k}$ is defined at position $\boldsymbol{r}$ and time t. There will be a particular choice of unit vector $\boldsymbol{n}$ for which the scalar flux density J will be a maximum (for given values of ΔA and δt), which will be called $\boldsymbol{n}_{\max}$. A vector-valued flux density $\boldsymbol{J}$ is defined so that

$$\boldsymbol{J} = J(\boldsymbol{n}_{\max}, \boldsymbol{r}, t)\, \boldsymbol{n}_{\max}, \tag{3.11}$$

that is, it points in the direction of the greatest flux density at position $\boldsymbol{r}$ and time t, and its magnitude, $J_{\max}$, is equal to the flux density in that direction. We shall see in the next subsection that $\boldsymbol{J}$ enables us to calculate the flux through an arbitrary surface element located at position $\boldsymbol{r}$.

3.3.4 Relating flux to the vector flux density

Having defined the vector flux density $\boldsymbol{J}$, we now consider how the flux Φ across a surface may be expressed in terms of $\boldsymbol{J}$. We start by discussing the flux across a small surface element of area ΔA, and proceed to considering the flux across an arbitrary surface.

Consider the flux $\Delta\Phi$ across a small element of area of size ΔA, with $\boldsymbol{n}$ being the unit vector normal to the surface. It will now be shown that if the angle between the normal vector $\boldsymbol{n}$ and the vector $\boldsymbol{n}_{\max}$ is θ, the flux is $J = J_{\max}\cos\theta$. For simplicity, we first consider the case where all of the particles are moving in the same direction (which coincides with the direction of the vector $\boldsymbol{n}_{\max}$). In Figure 3.5 the area ΔA is crossed by the same number of particles per unit time as ΔA_0, where ΔA_0 is a surface element perpendicular to $\boldsymbol{n}_{\max}$. The flux $\Delta\Phi$ through ΔA is therefore the same as that through ΔA_0, but the flux densities $J = \Delta\Phi/\Delta A$ and $J_{\max} = \Delta\Phi/\Delta A_0$ are different. The ratio of areas is $\Delta A_0/\Delta A = \cos\theta$, and

$$\frac{J}{J_{\max}} = \frac{\Delta A_0}{\Delta A}, \tag{3.12}$$

so

$$J = J_{\max}\cos\theta = J_{\max}\,\boldsymbol{n}_{\max} \cdot \boldsymbol{n} = \boldsymbol{J} \cdot \boldsymbol{n}. \tag{3.13}$$

Equation (3.13) relies upon the definition of the scalar product $\boldsymbol{a} \cdot \boldsymbol{b}$ of two vectors $\boldsymbol{a}$ and $\boldsymbol{b}$. If the angle between the directions of these vectors is θ, and their magnitudes are a and b, respectively, the scalar product is defined to be equal to $ab\cos\theta$, so $\boldsymbol{n} \cdot \boldsymbol{n}_{\max} = \cos\theta$.

See Block 0, Section 2.3.1.

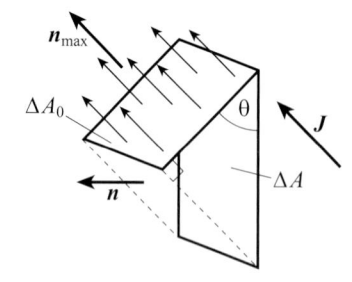

Figure 3.5 The flux passing through the two elements with areas ΔA and ΔA_0 is the same

Let us see how the flux $\Delta \Phi = J \Delta A$ can be expressed more elegantly in vector notation. We can introduce a vector surface element of area, $\Delta \boldsymbol{A} = \boldsymbol{n} \, \Delta A$, associated with a surface element of area ΔA with normal vector $\boldsymbol{n}$. The flux crossing the surface may then be written as $\Delta \Phi = (\boldsymbol{J} \cdot \boldsymbol{n}) \Delta A$, that is,

$$\Delta \Phi = \boldsymbol{J} \cdot \Delta \boldsymbol{A}. \tag{3.14}$$

This is a simple and general expression for the flux through a small surface element of area ΔA: this equation, and its integrated form equation (3.18), are the principal results of this subsection.

Let us consider how this general formula applies to the example shown in Figure 3.5. The vector $\Delta \boldsymbol{A}_0$ associated with the element of area ΔA_0 is in the same direction as the vector $\boldsymbol{J}$. When the normal to the surface is aligned with the direction of the vector flux density, the scalar flux density takes its maximum value $J_{\max}$, so equation (3.14) gives $\Delta \Phi = J_{\max} \Delta A_0$. The vector $\Delta \boldsymbol{A}$ associated with the element of area ΔA is at an angle θ to the direction of $\boldsymbol{J}$, so $\Delta \Phi = J_{\max} \boldsymbol{n}_{\max} \cdot \boldsymbol{n} \, \Delta A = J_{\max} \Delta A \cos \theta = J \Delta A$.

> In the case where the particles do not all move with the same velocity, equation (3.14) remains valid. This may be understood by considering the case where the particles move with M different velocities, which are labelled by an integer index taking values $k = 1, 2, \ldots, M$. The particles with the velocity labelled by k give rise to a flux density $\boldsymbol{J}_k$. Each of these gives a separate additive contribution to the flux $\Delta \Phi$:
>
> $$\Delta \Phi = \sum_{k=1}^{M} \boldsymbol{J}_k \cdot \Delta \boldsymbol{A} \ . \tag{3.15}$$
>
> This equation may also be written in the form $\Delta \Phi = \boldsymbol{J} \cdot \Delta \boldsymbol{A}$, which is equal to equation (3.14), if the total flux density $\boldsymbol{J}$ is given by
>
> $$\boldsymbol{J} = \sum_{k=1}^{M} \boldsymbol{J}_k. \tag{3.16}$$
>
> Any continuous distribution of velocities may be approached by taking the limit as $M \to \infty$, so equation (3.14) is valid in the general case.

Having obtained a general expression for the flux across an element of area in the form $\Delta \Phi = \boldsymbol{J} \cdot \Delta \boldsymbol{A}$, we now consider how to obtain the flux across a general surface. We divide this surface S into a large number of small vector elements $\Delta \boldsymbol{A}_i$, labelled by an index $i = 1, \ldots, M$. (See Figure 3.6, and note that the significance of the subscript label and its upper limit M are different from those in equation (3.16).)

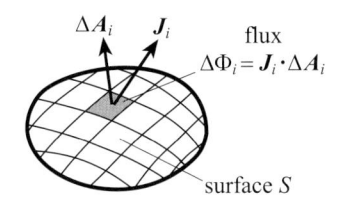

Figure 3.6 The flux across a surface S may be expressed as the sum of contributions $\Delta\Phi_i$ from small vector elements $\Delta\boldsymbol{A}_i$

The flux $\Delta\Phi_i$ across the vector element $\Delta\boldsymbol{A}_i$ is approximately $\boldsymbol{J}_i \cdot \Delta\boldsymbol{A}_i$, where $\boldsymbol{J}_i$ is the flux density evaluated at the centre of this element. The total flux is

$$\Phi = \sum_{i=1}^{M} \Delta\Phi_i \simeq \sum_{i=1}^{M} \boldsymbol{J}_i \cdot \Delta\boldsymbol{A}_i. \tag{3.17}$$

The accuracy of this approximation improves as we make the size of the elements $\Delta\boldsymbol{A}_i$ smaller (within the limits set by the discreteness of the particles discussed in Subsection 3.3.1, which we do not consider further). In the limit where the size of the elements $\Delta\boldsymbol{A}_i$ tends to zero, this sum becomes a surface integral, as discussed in Chapter 2 of Block 0. The flux is then given by

$$\Phi = \int_S d\boldsymbol{A} \cdot \boldsymbol{J}. \tag{3.18}$$

3.3.5 Defining flux density in one dimension

Sometimes we shall consider one-dimensional situations, so it will be helpful to consider the definition of flux density in one dimension.

In one dimension it is possible to define the flux Φ as the number of particles per unit time passing a given point. There is no area perpendicular to the direction of flow, so it is not possible to define a true flux density. However, it will be convenient to use the symbol J for the flux in one-dimensional problems, and this will be referred to as the flux density. The motivation for choosing this notation and terminology is that it makes the form of the *continuity equation* (which will be introduced in the next section) equivalent in one, two and three dimensions.

In one dimension, the flux density J is therefore defined as follows. Let δn be the number of particles passing the point x from left to right, minus the number passing from right to left, between times t and $t + \delta t$. This may be written as

$$\delta n = J(x,t)\,\delta t, \tag{3.19}$$

which defines the symbol J; that is, $J = \delta n/\delta t$. If the number of particles is sufficiently large, the quantity $J(x,t)$ is expected to be almost independent of the choice of δt over a very wide range of values (in the sense that the change δJ due to changing δt satisfies $\delta J/J \ll 1$).

3.3.6 Summary of concentration and flux density

Equations (3.20) and (3.21) below provide a concise summary of the important properties of concentration c and vector flux density $\boldsymbol{J}$. The number of particles N inside a volume V at time t is expressed in terms of the concentration c via a volume integral

$$N(t) = \int_V dV \, c(\boldsymbol{r},t). \tag{3.20}$$

The number of particles per unit time (the flux) Φ crossing a surface S at time t is expressed in terms of the flux density $\boldsymbol{J}$ via a surface integral

$$\Phi(t) = \int_S d\boldsymbol{A} \cdot \boldsymbol{J}(\boldsymbol{r},t). \tag{3.21}$$

Equations (3.20) and (3.21) could have been used as definitions of c and $\boldsymbol{J}$.

When defining the flux, we have a choice about which is the positive direction for counting the particles crossing the surface. For a closed surface there is normally an inside and an outside. We shall usually adopt the convention that the normal to such a surface points outwards.

A closed surface is one with no boundary, such as the surface of a sphere.

Exercise 3.9

(a) The concentration of particles is given by $c(x,y,z) = 1 + \frac{1}{10}x^2 + \frac{1}{4}xy^2$. What is the number of particles inside the unit cube defined by the conditions $0 < x < 1$, $0 < y < 1$, $0 < z < 1$?

If the concentration is measured in particles per unit volume, is the answer to this question meaningful? What about if the concentration is measured in mols per unit volume?

(b) The flux density is $\boldsymbol{J} = (2 + \frac{1}{10}z + \frac{1}{8}x^2)\mathbf{i} + 3\mathbf{j} + x^2\mathbf{k}$. What is the flux of particles across the side of this cube defined by the condition $x = 1$, with the normal to the surface in the direction of the unit vector $\mathbf{i}$?

Exercise 3.10

The concentration c in a three-dimensional region depends upon the distance r from the origin. Express the number of particles $N(R)$ inside a sphere of radius R in terms of an integral over the function $c(r)$. In the case where $c(r) = 1 + 1/r$, show that $N(R) = \frac{4}{3}\pi R^3 + 2\pi R^2$.

Exercise 3.11

Consider the situation where the flux density $\boldsymbol{J}$ is always directed away from the origin, and has a magnitude J which depends only upon the distance r from the origin. Write the outward flux $\Phi(R)$ across a spherical surface of radius R from the origin in terms of the flux density $J(r)$. Show that the flux is independent of radius if J is proportional to $1/r^2$.

3.4 The continuity equation

In the previous section, we introduced the concepts of concentration $c(\boldsymbol{r}, t)$ (describing the number of particles present in the vicinity of a point), and the flux density $\boldsymbol{J}(\boldsymbol{r}, t)$ (describing how these particles are moving). As the next step towards obtaining the diffusion equation, we are now going to obtain a partial differential equation, called the *continuity equation*, relating the flux density to the rate of change of concentration. The principle involved is quite simple: if the number of particles inside a small region is increasing, then the concentration increases. The net number of particles entering a region depends upon the balance of fluxes across the surface of the region, and if the region is very small, then this can be expressed in terms of the spatial derivatives of the flux density. We shall look at this first in one dimension, before turning to the three-dimensional case.

> Before starting the calculation, it may be helpful to say a little about the notation. You need not take in everything before proceeding to read the calculation, but you may find it useful to return to this comment if you find the notation confusing.
>
> Spatial increments will be written with an upper-case delta; for example, ΔV is a small volume element. The small increment in time will be written δt.
>
> The calculation uses three similar symbols, all representing relatively small changes in numbers of particles:
> - ΔN is the number of particles in a small volume element ΔV;
> - δN is the change in the number of particles in the volume element occurring in a short time δt;
> - δn is the number of particles passing through one surface of the volume element in time δt.
>
> Thus, changes preceded by δ are changes over a time interval δt, and may be divided by δt to obtain a rate of change.
>
> In order that the same definitions can be used in both the one-dimensional case and the three-dimensional case, in one dimension the 'volume' of the small element must be interpreted as its length, and the 'surfaces' are the two end points of the element.

3.4.1 The continuity equation in one dimension

In one dimension, the number of particles in the small interval $[x, x + \Delta x]$ is, by definition, $\Delta N = c(x,t) \, \Delta x$ (where $c(x,t)$ is the concentration). The change δN in ΔN in time δt is given by the net number of particles entering the interval by passing the point x from the left, minus the net number leaving the interval by passing $x + \Delta x$ to the right:

$$\delta N = \delta n(x,t) - \delta n(x + \Delta x, t), \tag{3.22}$$

where $\delta n(x,t)$ is the net number of particles passing the point x, from left to right, between t and $t + \delta t$. Expressing δN in terms of the flux density J, as defined in equation (3.19), we have

$$\delta N = [J(x,t) - J(x + \Delta x, t)] \, \delta t \tag{3.23}$$

(see Figure 3.7).

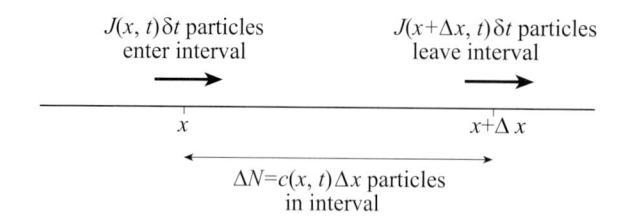

Figure 3.7 Illustrating the derivation of the continuity equation in one dimension: the number of particles within the interval is related to the concentration c, and the number of particles entering or leaving is related to the flux density J

The change in concentration occurring in time δt is

$$c(x, t + \delta t) - c(x, t) = \frac{\Delta N(x, t + \delta t) - \Delta N(x, t)}{\Delta x} = \frac{\delta N}{\Delta x}. \tag{3.24}$$

Dividing by δt and substituting from equation (3.23), we have

$$\frac{c(x, t + \delta t) - c(x, t)}{\delta t} = \frac{1}{\Delta x} \frac{\delta N}{\delta t}$$
$$= -\frac{J(x + \Delta x, t) - J(x, t)}{\Delta x}. \tag{3.25}$$

Taking the limits as $\Delta x \to 0$ and $\delta t \to 0$, then adding $\partial J/\partial x$ to both sides, we obtain

$$\frac{\partial c}{\partial t} + \frac{\partial J}{\partial x} = 0. \tag{3.26}$$

This equation is known as the *continuity equation*, relating flux density and concentration. It is a very general result. The only physical principle used in the derivation is that the particles move around without being created or destroyed: the particles are said to be *conserved*. This is a very common situation, and the continuity equation is therefore one of the fundamental equations of applied mathematics.

The detailed calculation presented above is rather cumbersome, but you should note that its essential elements are quite simple. The rate of change of concentration, $\partial c/\partial t$, must be equal to the difference in the rates at which particles enter and leave a small interval, divided by the length of the interval. The rates at which particles enter and leave the element are respectively $J(x, t)$ and $J(x + \Delta x, t)$, and their difference is approximately $-(\partial J/\partial x)\Delta x$. Dividing by Δx, we obtain the continuity equation $\partial c/\partial t = -\partial J/\partial x$.

Further insight into equation (3.26) can be gained by integrating it with respect to x on some interval (not necessarily small) of the real line $[a, b]$. This gives

$$\frac{\partial N}{\partial t} = J(a, t) - J(b, t), \tag{3.27}$$

where $N(t) = \int_a^b dx\, c(x, t)$ is the number of particles inside the interval $[a, b]$. Equation (3.27) is telling us that the rate of increase in $N(t)$ is equal to the flux density of particles entering the interval at $x = a$, namely $J(a, t)$, minus the flux density of particles leaving the interval at $x = b$, namely $J(b, t)$.

3.4.2 The continuity equation in three dimensions

In three dimensions, the derivation is similar. Consider the rate of change of concentration of particles in a small cuboidal element with sides Δx, Δy, Δz aligned with the Cartesian axes (see Figure 3.8).

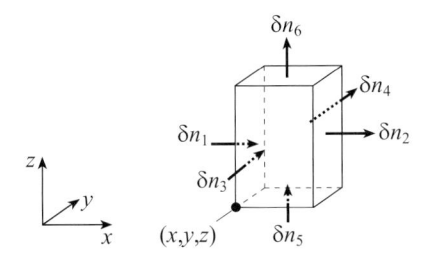

Figure 3.8 The rate of change of concentration within this volume element is determined by the total flux across its six surfaces

Let the number of particles inside this element of volume $\Delta V = \Delta x \, \Delta y \, \Delta z$ be ΔN, and let the change in ΔN in time δt be δN. The quantity δN is equal to the sum of the net numbers of particles entering through each of the six sides in a time interval of length δt (see Figure 3.8). Consider one of these sides, a rectangular surface with a constant value of the coordinate, x say (with the other coordinates in the intervals $[y, y + \Delta y]$ and $[z, z + \Delta z]$). The area of this surface element has magnitude $\Delta A = \Delta y \, \Delta z$, and the inward normal to this surface element is $\boldsymbol{n} = \mathbf{i}$, where $\mathbf{i}$ is the unit vector aligned with the x-axis. (In this calculation, we choose not to adopt the usual convention that all elements of area are directed outwards from a closed surface; instead, for convenience, we take them to be aligned with the unit vectors $\mathbf{i}$, $\mathbf{j}$ and $\mathbf{k}$.) The net number of particles passing through this small surface element in the direction of increasing x during time δt is

$$\delta n_1 = \boldsymbol{J} \cdot \Delta \boldsymbol{A} \delta t, \tag{3.28}$$

where $\boldsymbol{J} = J_1 \mathbf{i} + J_2 \mathbf{j} + J_3 \mathbf{k}$ is evaluated at the centre of the element, at position $(x, y + \tfrac{1}{2}\Delta y, z + \tfrac{1}{2}\Delta z)$. The dot product $\boldsymbol{J} \cdot \Delta \boldsymbol{A}$ is simply $J_1 \Delta y \, \Delta z$, because the direction of $\Delta \boldsymbol{A} = \Delta y \, \Delta z \, \mathbf{i}$ is aligned with the x-axis. The net number of particles passing the surface in time δt is therefore

$$\delta n_1 = J_1(x, y + \tfrac{1}{2}\Delta y, z + \tfrac{1}{2}\Delta z, t) \, \Delta y \, \Delta z \, \delta t. \tag{3.29}$$

We can now expand J_1 as a Taylor series about (x, y, z, t):

$$J_1(x, y + \tfrac{1}{2}\Delta y, z + \tfrac{1}{2}\Delta z, t) = J_1(x, y, z, t) + \tfrac{1}{2}\Delta y \frac{\partial J_1}{\partial y} + \tfrac{1}{2}\Delta z \frac{\partial J_1}{\partial z} + \cdots \tag{3.30}$$

(where the partial derivatives are all evaluated at (x, y, z, t)). Terms proportional to higher powers of Δy and Δz will be generated, but these terms may be neglected in the limit $\Delta y \to 0$, $\Delta z \to 0$.

Similarly, the formula for the net number of particles leaving via the opposite face, at $x + \Delta x$, is

$$\delta n_2 = J_1(x + \Delta x, y + \tfrac{1}{2}\Delta y, z + \tfrac{1}{2}\Delta z, t) \, \Delta y \, \Delta z \, \delta t. \tag{3.31}$$

Note that δn_2 makes a contribution to δN which differs in sign from δn_1 in equation (3.29), because particles crossing in the direction of increasing x across that face are leaving the volume element. The expression for δn_2 differs from δn_1 only in that J_1 is evaluated at position $(x + \Delta x, y + \tfrac{1}{2}\Delta y, z + \tfrac{1}{2}\Delta z)$. Expanding the value of J_1 about (x, y, z, t) at this position gives an additional term of first order in Δx, namely $\Delta x \, \partial J_1 / \partial x$, as well as the same

Δy and Δz terms as before. The difference between the numbers of particles crossing these opposite faces shown in Figure 3.8 is thus

$$\delta n_1 - \delta n_2 = \left[J_1(x, y + \tfrac{1}{2}\Delta y, z + \tfrac{1}{2}\Delta z, t)\right.$$
$$\left. - J_1(x + \Delta x, y + \tfrac{1}{2}\Delta y, z + \tfrac{1}{2}\Delta z, t)\right] \Delta y \, \Delta z \, \delta t$$
$$\simeq -\frac{\partial J_1}{\partial x}(x, y, z, t) \Delta x \, \Delta y \, \Delta z \, \delta t. \tag{3.32}$$

Repeating this calculation for the differences $\delta n_3 - \delta n_4$ and $\delta n_5 - \delta n_6$ gives similar expressions, containing $\partial J_2/\partial y$ and $\partial J_3/\partial z$, respectively. Neglecting terms of higher order in the dimensions of the box (i.e. terms of higher order in Δx, Δy and Δz), the total gain in the number of particles within the volume element in time δt is the sum of six contributions, corresponding to the six fluxes illustrated in Figure 3.8:

$$\delta N = \delta n_1 - \delta n_2 + \delta n_3 - \delta n_4 + \delta n_5 - \delta n_6$$
$$\simeq -\left[\frac{\partial J_1}{\partial x} + \frac{\partial J_2}{\partial y} + \frac{\partial J_3}{\partial z}\right] \Delta x \, \Delta y \, \Delta z \, \delta t. \tag{3.33}$$

Dividing δN by $\Delta V = \Delta x \, \Delta y \, \Delta z$ gives the change in concentration in time δt, so, neglecting terms of higher order in Δx, Δy and Δz, dividing by δt and taking limits as $\delta t \to 0$ and $\Delta V \to 0$, we have

$$\frac{\partial c}{\partial t} + \left(\frac{\partial J_1}{\partial x} + \frac{\partial J_2}{\partial y} + \frac{\partial J_3}{\partial z}\right) = 0. \tag{3.34}$$

The combination of derivatives in the bracket occurs frequently in discussions of vector fields, and we saw in Block 0, Chapter 2 that it is called the *divergence* of the vector field $\boldsymbol{J}$. There are two commonly used abbreviated notations for the divergence of a vector field $\boldsymbol{F}$:

$$\left(\frac{\partial F_1}{\partial x} + \frac{\partial F_2}{\partial y} + \frac{\partial F_3}{\partial z}\right) = \boldsymbol{\nabla} \cdot \boldsymbol{F} = \mathrm{div}\,\boldsymbol{F}. \tag{3.35}$$

The dot product notation in equation (3.35) arises from regarding $\boldsymbol{\nabla}$ as a vector operator with components which are differential operators:

$$\boldsymbol{\nabla} = \mathbf{i}\frac{\partial}{\partial x} + \mathbf{j}\frac{\partial}{\partial y} + \mathbf{k}\frac{\partial}{\partial z}. \tag{3.36}$$

The divergence $\boldsymbol{\nabla} \cdot \boldsymbol{F}$ is the scalar product of this vector differential operator and the vector field $\boldsymbol{F}$. With these notational conventions, the *three-dimensional continuity equation* takes the form

$$\frac{\partial c}{\partial t} + \boldsymbol{\nabla} \cdot \boldsymbol{J} = 0. \tag{3.37}$$

A commonly occurring situation is that of a steady flow, i.e. with the concentration independent of time, so that $\partial c/\partial t = 0$ at each point in space. In this case the vector flux density $\boldsymbol{J}$ has zero divergence, i.e. $\boldsymbol{\nabla} \cdot \boldsymbol{J} = 0$ at each point. This situation is also referred to as the *steady state*.

The continuity equation is also important in describing the flow of fluids, where it expresses the fact that the total mass of the fluid remains constant. The derivation is similar to that for concentration. The total mass M in a volume V is an integral of the mass density ρ_m, which may be a function of both position and time:

$$M = \int_V dV \, \rho_\mathrm{m}(\boldsymbol{r}, t). \tag{3.38}$$

This is analogous to equation (3.20). All of the fluid particles in the neighbourhood of a given point are understood to be moving with a velocity $\boldsymbol{v}$, which may also depend upon both position $\boldsymbol{r}$ and time t. Using arguments similar to those of Subsection 3.3.4, it can be shown that the mass of fluid

3.4 The continuity equation

passing through a surface element $\Delta \boldsymbol{A}$ in time δt is $\delta M = \rho_{\mathrm{m}} \boldsymbol{v} \cdot \Delta \boldsymbol{A}\, \delta t$. A vector flux density for the flow of mass is defined, $\boldsymbol{J}_{\mathrm{mass}} = \rho_{\mathrm{m}} \boldsymbol{v}$, such that the rate of change of the mass of fluid inside volume V is given by integrating this flux density over the surface S which is the boundary of V:

$$\frac{dM}{dt} = -\int_S d\boldsymbol{A} \cdot \rho_{\mathrm{m}} \boldsymbol{v}. \tag{3.39}$$

This is analogous to equation (3.21). The minus sign in equation (3.39) occurs because we have used the convention that $d\boldsymbol{A}$ points in the direction of the outward normal to the surface, and M decreases as mass flows in this direction. All of the arguments used to obtain the continuity equation apply equally well to equations (3.38) and (3.39). Replacing c by ρ_{m} and $\boldsymbol{J}$ by $\boldsymbol{J}_{\mathrm{mass}} = \rho_{\mathrm{m}} \boldsymbol{v}$ shows that, for fluid flow, the continuity equation is of the form

$$\frac{\partial \rho_{\mathrm{m}}}{\partial t} + \boldsymbol{\nabla} \cdot (\rho_{\mathrm{m}} \boldsymbol{v}) = 0. \tag{3.40}$$

Exercise 3.12

If a is a scalar field and $\boldsymbol{F}$ is a vector field, show that

$$\boldsymbol{\nabla} \cdot (a\boldsymbol{F}) = a \boldsymbol{\nabla} \cdot \boldsymbol{F} + \boldsymbol{F} \cdot \boldsymbol{\nabla} a.$$

Exercise 3.13

In most liquids, the mass density, ρ_{m}, can be treated as a constant (i.e. as independent of $\boldsymbol{r}$ and t). Show that the velocity $\boldsymbol{v}(\boldsymbol{r}, t)$ of such a liquid satisfies $\boldsymbol{\nabla} \cdot \boldsymbol{v} = 0$.

Exercise 3.14

Consider the time-independent vector fields

(a) $\boldsymbol{J} = x\mathbf{i} + y\mathbf{j} + z\mathbf{k}$,

(b) $\boldsymbol{J} = y\mathbf{i} - x\mathbf{j} + A\mathbf{k}$,

(c) $\boldsymbol{J} = (Ax + By)\mathbf{i} + (Cx + Dy)\mathbf{j}$,

where A, B, C and D are constants. In each case, evaluate $\boldsymbol{\nabla} \cdot \boldsymbol{J}$, and determine which of these vector fields can represent the flux density of conserved particles in a steady flow. [Hint: Use the continuity equation (3.37).]

Show that in case (c) this is possible only if $A + D = 0$.

3.4.3 Relation with Gauss's theorem

The continuity equation is very closely related to Gauss's theorem. We shall demonstrate this relationship by combining the continuity equation (3.37) with equation (3.18) to deduce Gauss's theorem.

Gauss's theorem was discussed in Block 0, Chapter 2.

Let S be a closed surface, containing a volume V. Using equation (3.7) and differentiating with respect to t, the rate of change of the number of particles inside V is

$$\frac{dN}{dt} = \int_V dV \frac{\partial c}{\partial t}(\boldsymbol{r}, t). \tag{3.41}$$

However, the rate, $-dN/dt$, at which particles are leaving the volume V is (by definition) equal to the flux, Φ, across the surface (in the outward direction). Using equation (3.18) for the flux (with the convention that $d\boldsymbol{A}$

points in the direction of the outward normal to the surface), and equating the result with equation (3.41), we have

$$\Phi = \int_S d\boldsymbol{A} \cdot \boldsymbol{J} = -\frac{dN}{dt} = -\int_V dV\, \frac{\partial c}{\partial t}. \tag{3.42}$$

Equating the two integrals in equation (3.42), and using the continuity equation (3.37) to substitute for $\partial c/\partial t$, we have

$$\int_S d\boldsymbol{A} \cdot \boldsymbol{J} = \int_V dV\, \boldsymbol{\nabla} \cdot \boldsymbol{J}. \tag{3.43}$$

This equation was obtained using physical arguments in which the vector field $\boldsymbol{J}(\boldsymbol{r},t)$ was interpreted as a flux density. There is, however, no restriction on the choice of the vector field, and this equation must be true for any closed surface which is sufficiently smooth that the divergence of the vector field exists and the integrals are well-defined. We conclude that the following identity therefore holds for any closed surface S containing a volume V in which a differentiable vector field $\boldsymbol{F}$ is defined:

$$\int_S d\boldsymbol{A} \cdot \boldsymbol{F} = \int_V dV\, \boldsymbol{\nabla} \cdot \boldsymbol{F}. \tag{3.44}$$

This is *Gauss's theorem*, a fundamental result in the theory of vector calculus, which extends the concepts of differentiation and integration to vectors. It enables integrals over surfaces to be expressed as integrals over volumes, and vice versa.

Exercise 3.15

Consider the vector field $\boldsymbol{F} = (x + y^2)\mathbf{i} + (2y^2 - x^3)\mathbf{j}$.

(a) Calculate the surface integral

$$\Phi_S = \int_S d\boldsymbol{A} \cdot \boldsymbol{F},$$

where S is the surface of a cube with its six faces lying in the planes $x = 0$, $x = 1$, $y = 0$, $y = 1$, $z = 0$, $z = 1$.

(b) Calculate $\boldsymbol{\nabla} \cdot \boldsymbol{F}$, and the volume integral

$$\Phi_V = \int_V dV\, \boldsymbol{\nabla} \cdot \boldsymbol{F},$$

where V is the cube with surface S. Verify that Gauss's theorem is satisfied.

3.5 The diffusion equation

The diffusion equation can be obtained from the continuity equation (3.37), together with a simple assumption about the relationship between flux density and concentration. This assumption is often known as *Fick's law*.

> *Fick's law*: The flux density is proportional to the concentration gradient, with flow proceeding from high to low concentration. In vector notation
> $$\boldsymbol{J} = -D\,\boldsymbol{\nabla}c = -D\left(\frac{\partial c}{\partial x}\mathbf{i} + \frac{\partial c}{\partial y}\mathbf{j} + \frac{\partial c}{\partial z}\mathbf{k}\right), \quad (3.45)$$
> where D is a positive constant known as the *diffusion constant*.

Adolf E. Fick (1829–1901) was a German physiologist, best known in physiology for his work on cardiac output (1870), making possible the evaluation of respiratory exchange, i.e. the delivery of oxygen to bodily tissues. He is also credited with developing the first contact lens in 1887. He formulated his law of diffusion in 1855.

Fick's law was originally proposed as an empirical law. The negative sign is present because particles are expected to move against the concentration gradient, away from regions of higher concentration (see Figure 3.9). Fick's law is a very natural guess for a relationship between the vector field $\boldsymbol{J}$ and the scalar field c, because taking the gradient is the simplest way to construct a vector field from a scalar field.

An empirical law is one supported by experimental observation, rather than being deduced from fundamental principles.

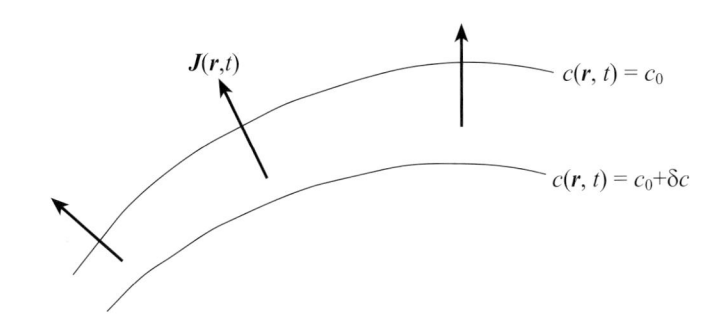

Figure 3.9 Illustrating Fick's law: the flux density of diffusing particles $\boldsymbol{J}$ points from regions of higher concentration to regions of lower concentration, and is proportional to the concentration gradient. Here, $\delta c > 0$.

To understand how this leads to the diffusion equation, consider first the one-dimensional case. Here the flux density is given by the one-dimensional version of Fick's law
$$J = -D\frac{\partial c}{\partial x}. \quad (3.46)$$

Substituting this into the one-dimensional version of the continuity equation (3.26) leads directly to the one-dimensional diffusion equation (3.1):
$$\frac{\partial c}{\partial t} = -\frac{\partial J}{\partial x} = -\frac{\partial}{\partial x}\left(-D\frac{\partial c}{\partial x}\right) = D\frac{\partial^2 c}{\partial x^2}. \quad (3.47)$$

We used the fact that D is a constant in the final equality.

In three dimensions, combining Fick's law and the continuity equation (3.37) gives
$$\frac{\partial c}{\partial t} = -\boldsymbol{\nabla}\cdot\boldsymbol{J} = -\boldsymbol{\nabla}\cdot(-D\,\boldsymbol{\nabla}c). \quad (3.48)$$

The following holds for any scalar field f:

$$\nabla \cdot (\nabla f) = \frac{\partial}{\partial x}\left(\frac{\partial f}{\partial x}\right) + \frac{\partial}{\partial y}\left(\frac{\partial f}{\partial y}\right) + \frac{\partial}{\partial z}\left(\frac{\partial f}{\partial z}\right) = \nabla^2 f. \tag{3.49}$$

Using this relation in equation (3.48), and noting that D is a constant coefficient, we obtain the three-dimensional diffusion equation (3.2)

$$\frac{\partial c}{\partial t} = D \nabla^2 c. \tag{3.50}$$

If the region in which diffusion occurs is finite, *boundary conditions* are required, which describe the behaviour of the concentration c at the boundary of the region. These are determined by physical principles, rather than mathematical reasoning. The most usual case is that of diffusion in a finite region with an impermeable boundary (i.e. no particles are able to enter or leave the region). An example would be diffusion of a substance through a liquid contained in a glass beaker: the substance remains confined to the liquid. The mathematical expression of this fact is the statement that, for all positions on the boundary, there is no flux crossing the boundary: i.e. $\boldsymbol{J}(\boldsymbol{r}) \cdot \boldsymbol{n} = 0$, where $\boldsymbol{n}$ is a unit vector normal to the surface, for all positions $\boldsymbol{r}$ on the boundary, and for all times. Using Fick's law, this condition is expressed in terms of the concentration:

$$\boldsymbol{n} \cdot \nabla c = 0 \tag{3.51}$$

for all points on the boundary and for all times t larger than the initial time (which is usually taken to be zero). This boundary condition is known as a *Neumann boundary condition*.

> Another physical requirement on the solution of the diffusion equation is that as c represents a concentration of particles, this quantity must always remain non-negative. This constraint does not restrict the solution, in that if $c(\boldsymbol{r}, t)$ is initially everywhere non-negative, this property remains true of the solution at later times. This property, that non-negative initial conditions lead to non-negative solutions, will become apparent from a general solution given later, in Chapter 4.

The derivation of the diffusion equation given in this chapter was based upon an assumption about the macroscopic behaviour of diffusing substances, namely Fick's law. It is desirable to have a derivation based upon microscopic principles; this will be given in Chapter 6.

Exercise 3.16

Consider the function $c(x,t) = \exp(-x/t)/t$, which represents concentration as a function of time, in the region $x \geq 0$. The particles are unable to enter the region $x < 0$, so the flux density is zero at $x = 0$. Show that the number of particles is conserved. Use the continuity equation to deduce the flux density $J(x,t)$. Does $c(x,t)$ satisfy Fick's law?

Exercise 3.17

Consider the scalar function $f(r) = 1/r$, where r is the distance from the origin. What is the gradient of this function? By considering the flux of this gradient vector (as defined by equation (3.18) with $\boldsymbol{J} = \nabla f$) over the surface bounding two spheres of different radii centred on the origin, show that $\nabla^2 f = 0$ everywhere except at $r = 0$. [Hint: Evaluating the integral in equation (3.18) is straightforward, because $\boldsymbol{J} \cdot \boldsymbol{n}$ is constant across the surface of a sphere.]

Exercise 3.18

Consider a three-dimensional region where the concentration at time t depends only upon the distance r from the origin, and is given by the function $c(r,t)$. The vector flux density $\mathbf{J}$ must point in a radial direction, and at time t it can depend only on r: let the scalar flux density in the radially outward direction be $J(r,t)$.

(a) By considering the flux of particles across the surface of a sphere of radius R centred on the origin, and relating this to the decrease of the number of particles within the sphere, show that $J(r,t)$ can be expressed in terms of $c(r,t)$ via the integral

$$J(r,t) = -\frac{1}{r^2} \int_0^r ds \, s^2 \frac{\partial c}{\partial t}(s,t).$$

(b) If the concentration obeys the diffusion equation (when expressed as a function of x, y, z and t), use Fick's law to deduce that $c(r,t)$ satisfies

$$\frac{\partial c}{\partial t} = D \frac{1}{r^2} \frac{\partial}{\partial r}\left(r^2 \frac{\partial c}{\partial r}\right).$$

3.6 The heat equation

The diffusion equation was originally known as the *heat equation*. Let us now look at the connection.

Heat is a form of energy, consisting of random microscopic motion of atoms. The *temperature* θ of a material is a measure of the amount of heat energy per unit volume, i.e. the concentration of heat energy. In most substances, the temperature measured by a thermometer placed in contact with the substance may be approximated by a linear function of the concentration of heat energy q, which we write in the form

$$\theta = \theta_0 + \frac{1}{C\rho_{\mathrm{m}}}(q - q_0). \tag{3.52}$$

This is an approximate relationship, which is justified by experimental observation and by physical theory. Here ρ_{m} is the density of the material, C is the *mass-specific heat capacity* of the material, and θ_0, q_0 are two other constants. This expression is a good approximation when $q - q_0$ and $\theta - \theta_0$ are small. It is not necessary to be familiar with specific heat capacities; the important point is that θ and q are linearly related.

Exercise 3.19

(a) What is the SI unit of q?

(b) Roughly speaking, the temperature in Kelvins is the temperature in centigrade plus 273.15, the latter number being chosen so that zero Kelvins is the lowest attainable temperature ('absolute zero'). What is the SI unit of mass-specific heat capacity?

The SI unit of temperature is the Kelvin, symbol K.

The amount of heat energy is conserved (in all of the processes that we shall consider), and q therefore obeys a continuity equation.

Heat energy can move through a substance by *thermal conduction*, causing the temperature in different parts of the substance to change. You will become aware of thermal conduction if you hold a poker in a fire: eventually you will be forced to drop the poker when the heat of the fire reaches the end you are holding. Thermal conduction is closely related to diffusion: it can be thought of as diffusion of heat energy. The flux density of heat energy $\boldsymbol{J}$ is assumed to be proportional to the temperature gradient, so

$$\boldsymbol{J} = -\kappa \boldsymbol{\nabla} \theta, \tag{3.53}$$

where the positive constant κ is the *thermal conductivity*. This equation is known as *Fourier's law of heat conduction*. Like Fick's law, it was an empirical law, justified by its success in explaining experimental observations. Note that the Fourier and Fick laws are of the same form, relating a flux density to a gradient. In Fourier's law, the negative sign occurs because (in agreement with everyday experience) heat energy flows away from hotter regions, causing temperature differences to decrease.

It was proposed by J. B. Fourier in 1807.

> Equation (3.53) applies only in the case where the material is *homogeneous* (has the same properties at all positions) and *isotropic* (the material properties have no preferred direction). Most materials satisfy these criteria.
>
> Exceptions include materials with a chemical composition which depends upon position (which would be inhomogeneous), or materials with fibres aligned along a given direction, such as some carbon-fibre composites (which would be anisotropic).
>
> The derivation of equation (3.53) from fundamental laws of physics is problematic, and for some apparently reasonable mathematical models of heat conduction it has been shown to fail entirely. However, for most isotropic and homogeneous materials at moderate temperatures, Fourier's law does give an accurate description of heat conduction.

We can obtain the heat/diffusion equation from equations (3.52) and (3.53). We can differentiate equation (3.52) to relate the time derivative of temperature to that of q:

$$\frac{\partial \theta}{\partial t} = \frac{1}{C \rho_\mathrm{m}} \frac{\partial q}{\partial t}. \tag{3.54}$$

Now the continuity equation for q (i.e. equation (3.37) with c replaced by q), together with equation (3.53), gives

$$\frac{\partial q}{\partial t} = -\boldsymbol{\nabla} \cdot \boldsymbol{J} = \kappa \nabla^2 \theta. \tag{3.55}$$

Combining the previous two equations, we see that the temperature therefore satisfies the equation

$$\frac{\partial \theta}{\partial t} = \frac{\kappa}{C \rho_\mathrm{m}} \nabla^2 \theta, \tag{3.56}$$

which is known as the *heat equation*. It is identical in form to the diffusion equation. The constant $\kappa/C\rho_\mathrm{m}$ is called the *thermal diffusivity*.

Thermal diffusivity has the same dimensions as a diffusion constant, and will be given the same symbol D (it will be clear from context when we are dealing with a thermal diffusivity).

> At the time when Fourier first formulated his law, scientists understood heat to be an invisible fluid, called 'caloric'. Temperature was thought of as being the pressure of this fluid, and Fourier's law then states that the rate of flow of heat is proportional to the pressure gradient. This picture can be very useful in developing intuition about problems involving the flow of heat. The *caloric model* had to be abandoned when it was discovered that heat could be transformed into mechanical energy. The modern understanding is that heat

is a form of energy, in which the energy is contained in the random motion of atoms.

Exercise 3.20

(a) The density, mass-specific heat capacity and thermal conductivity of air (dry, at 20°C and standard atmospheric pressure) are, respectively, $\rho_m = 1.29 \text{ kg m}^{-3}$, $C = 1.01 \times 10^3 \text{ J kg}^{-1} \text{ K}^{-1}$ and $\kappa = 2.40 \times 10^{-2} \text{ J m}^{-1} \text{ s}^{-1} \text{ K}^{-1}$. What is the thermal diffusivity, D, of air?

(b) The diffusion constants (for molecular diffusion) for water vapour and carbon dioxide in air are, respectively, $D_{H_2O} = 2.12 \times 10^{-5} \text{ m}^2 \text{ s}^{-1}$ and $D_{CO_2} = 1.29 \times 10^{-5} \text{ m}^2 \text{ s}^{-1}$. Does this suggest anything about the relationship between diffusion of heat and diffusion of molecules?

3.7 A solution of the diffusion equation

One particularly instructive solution of the diffusion equation describes what happens when N particles are initially concentrated at a single point, and then diffuse in an infinite region. We shall consider only the one-dimensional case for the moment; extension to two and three dimensions is considered briefly in Exercise 3.24, and in much more detail in Chapter 4. A suitable solution of the one-dimensional diffusion equation (3.1), with c replacing f, is a Gaussian function of the form

$$c(x,t) = \frac{N}{\sqrt{4\pi Dt}} \exp(-x^2/4Dt). \tag{3.57}$$

It can be verified by substitution that this is a solution of the diffusion equation (3.1) when $t > 0$ from above. As $t \to 0$, the solution is concentrated at $x = 0$ (in the sense that for any point other than $x = 0$, the solution approaches zero as $t \to 0$). This solution is illustrated in Figure 3.10.

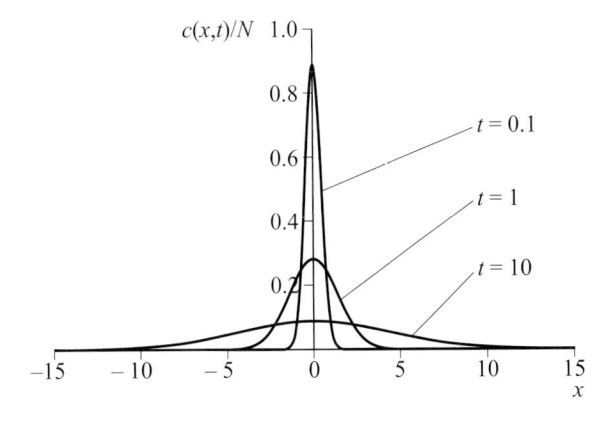

Figure 3.10 A Gaussian solution of the one-dimensional diffusion equation, in which N particles are initially all at $x = 0$. In this figure, we set $D = 1$.

Exercise 3.21

Check (by substitution) that (3.57) satisfies the diffusion equation (3.1) (with f replaced by c) when $t > 0$.

Exercise 3.22

If $c(x,t)$ satisfies the diffusion equation, show that $c(x - x_0, t - t_0)$ also satisfies the diffusion equation (where x_0 and t_0 are real constants). This means that solutions can be *translated* (moved) in both space and time.

Because the diffusion equation is linear, any constant multiple of a solution is also a solution. So, if we divide $c(\mathbf{r},t)$ by the total number of particles N to define

$$\rho(\mathbf{r},t) = c(\mathbf{r},t)/N, \tag{3.58}$$

then $\rho(\mathbf{r},t)$ will also be a solution to the same diffusion equation. We see from equation (3.7) that such a solution satisfies

$$\int_V dV \, \rho(\mathbf{r},t) = 1, \tag{3.59}$$

where V is the *full* volume containing *all* N particles. Any solution of the diffusion equation with this property will be called *normalised*. We can interpret a normalised solution as giving the probability density $\rho(\mathbf{r},t)$ for a particle to be located in the vicinity of $\mathbf{r}$ at time t (hence our use of the symbol ρ). The concentration is obtained by multiplying the normalised solution by the total number of particles present, N.

Thus, from equations (3.57) and (3.58), the normalised solution in one dimension is

$$\rho(x,t) = \frac{1}{\sqrt{4\pi Dt}} \exp(-x^2/4Dt), \tag{3.60}$$

which can be treated as a probability density. In the following exercise, we consider one of its moments.

Exercise 3.23

Confirm by direct integration that equation (3.60) is normalised, and that it satisfies

$$\int_{-\infty}^{\infty} dx \, x^2 \rho(x,t) = 2Dt.$$

Given that a normalised solution may be interpreted as a probability density, this exercise shows that the mean-squared distance of particles from their starting point is proportional to time, i.e.

$$\langle x^2 \rangle = 2Dt. \tag{3.61}$$

This should be compared with equation (2.18), which is a characteristic property of random walks. This is evidence that the underlying physical mechanism for diffusion is that the particles make random walks. In Chapter 6, the diffusion equation will be derived from the random walks exhibited by individual particles.

Exercise 3.24

(a) Show that a solution of the two-dimensional diffusion equation can be obtained as a product of two solutions of the one-dimensional equation.

(b) Use this approach to obtain the solution of the two-dimensional diffusion equation obtained in Exercise 3.2:
$$\rho(x,y,t) = \frac{C}{t} \exp[-(x^2 + y^2)/4Dt].$$
Determine the constant C which makes this a normalised solution.

(c) For the normalised solution in part (b), what is the mean-squared value of the distance to which the particles diffuse from the origin, $\langle r^2 \rangle$, where $r = \sqrt{x^2 + y^2}$? [Hint: You can use polar coordinates for the two-dimensional integral, but it is easier if you note that the double integral factorises.]

(d) What do you think $\langle r^2 \rangle$ will be in three dimensions? (In this case $r = \sqrt{x^2 + y^2 + z^2}$.)

3.8 Summary and outcomes

The principle ideas discussed in this chapter can be summarised as follows. The distribution in space of the quantity of a substance is described by its concentration c, and the movement of substances is described by the flux density $\boldsymbol{J}$. We showed that if the substance is not created or destroyed (that is, if it is conserved), then c and $\boldsymbol{J}$ are related by the continuity equation. We considered a particular type of particle-conserving process, called diffusion, in which substances move due to the random movement of individual atoms. We showed that a natural assumption about the flux density in diffusion, namely Fick's law, leads to the diffusion equation. We found that diffusion from a point source has the property that the mean-squared distance travelled is proportional to time, which indicates a connection between random walks and diffusion. The same concepts are applicable to heat flow, and the diffusion equation is also known as the heat equation.

After reading this chapter, you should:

- be aware of the macroscopic signs of diffusion, and of its microscopic physical mechanism;
- understand the definitions of the concentration $c(\boldsymbol{r},t)$ and the flux density $\boldsymbol{J}(\boldsymbol{r},t)$ — you should be able use these quantities to calculate the number of particles in a volume and the flux across a surface;
- be able to calculate simple surface and volume integrals, and to relate these using Gauss's theorem where appropriate;
- understand the continuity equation, and its role in deriving the diffusion equation;
- understand Fick's law of diffusion;
- understand the relation between the diffusion equation and the flow of heat;
- be aware of the Gaussian solution of the diffusion equation, and of its connections with the probability distribution for a random walk.

3.9 Further Exercises

These are harder exercises, which will challenge your understanding of the material.

Exercise 3.25

Assuming particle conservation, what is the boundary condition on the concentration at the smooth interface of two regions, numbered 1 and 2, with different diffusion constants, D_1 and D_2?

[Hint: The flux density across the boundary is the same on either side. Express the flux density on either side in terms of Fick's law, in order to give a condition which must be satisfied by the concentration c.]

Exercise 3.26

Two layers of material are bonded together. Their thicknesses and thermal conductivities are, respectively, L_1, κ_1 for material 1, and L_2, κ_2 for material 2. The double layer is used to separate two fluids, one at temperature θ_1, the other at θ_2, with material 1 in contact with the liquid at temperature θ_1. What is the temperature θ_i of the interface between the materials when a steady flow of heat has been established? What is the flux density of heat energy? [Hint: Consider one-dimensional heat flow in the direction perpendicular to the boundaries of the layers.]

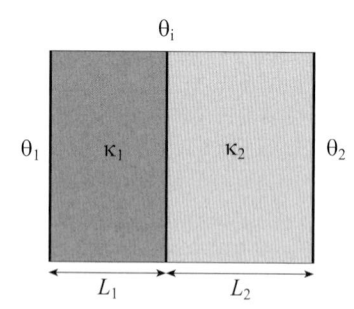

Figure 3.11 Illustrating Exercise 3.26: two layers of solid material separate two liquids at temperatures θ_1 and θ_2

Exercise 3.27

(a) Pipes carrying hot water are often surrounded by a layer of insulation in order to minimise heat loss. Consider a layer of insulation around a pipe, occupying the region bounded by two cylinders, with inner radius r_1 and outer radius r_2, which are kept at constant temperatures θ_1 and θ_2, respectively. (The cross-section is shown in Figure 3.12.) Assuming steady heat flow, show that the temperature at radius r (with $r_1 \leq r \leq r_2$) is

$$\theta(r) = \theta_1 + \frac{\theta_2 - \theta_1}{\ln(r_2/r_1)} \ln\left(\frac{r}{r_1}\right).$$

(b) What is the flux of heat per unit length of pipe? Does doubling the thickness of insulation halve the rate at which heat is lost?

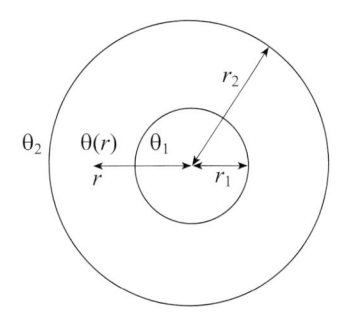

Figure 3.12 Illustrating Exercise 3.27: a pipe is surrounded by a layer of insulation

[Hints: This problem has two simplifying features: it involves a steady state, and there is a high degree of symmetry, so the temperature depends upon only one coordinate, r.

Consider a thin cylindrical shell between radii r and $r + \delta r$. Because the temperature is steady, $\partial\theta/\partial t = 0$, and the flux of heat entering the inner surface ar r is balanced by the flux leaving the shell at $r + \delta r$. The heat flux across any cylindrical surface is therefore constant. This flux is proportional to the temperature gradient, $d\theta/dr$, and to the area, which is proportional to r, so $r\, d\theta/dr = \text{constant}$.]

Exercise 3.28

Consider the case where particles undergo diffusion in a fluid which is itself in motion with velocity $\boldsymbol{v}(\boldsymbol{r}, t)$. (An example is where a pollutant enters a water system: it spreads by diffusion and also because it is being carried along by the flow.) Assume that the flux density due to the flow of the fluid can be added to that which results from diffusion. Show that the flux density due to the fluid flow is $\boldsymbol{J}_{\text{flow}} = c\boldsymbol{v}$, where c is the concentration. Using the continuity equation, show that the equation for c is

$$\frac{\partial c}{\partial t} = D\nabla^2 c - \boldsymbol{v} \cdot \boldsymbol{\nabla} c - (\boldsymbol{\nabla} \cdot \boldsymbol{v})c.$$

This is sometimes called the *diffusion–advection equation*.

Exercise 3.29

(a) Consider the one-dimensional diffusion–advection equation where the fluid flow is steady, so that the velocity $v(x)$ does not depend upon t. Consider the case where the particles are confined to the region $x \geq 0$, and the concentration $c(x)$ approaches zero as $x \to \infty$. Show that a steady-state solution, where the concentration does not depend upon t, takes the form

$$c(x) = A\exp\left[\frac{1}{D}\int_0^x dy\, v(y)\right],$$

where A is a constant.

(b) If the fluid contains N particles, how might the constant A be determined?

(c) What is the form of the solution in the cases where (i) $v(x) = -\lambda$ and (ii) $v(x) = -\mu x$ (where λ and μ are constants)?

3.10 Appendix: observing diffusion in a simple experiment

If you would like to try an experiment to see diffusion for yourself, it is quite easily observed using a dye (such as food colouring) dissolved in water. It is important to avoid any macroscopic motion of the water. Figure 3.13 shows some photographs illustrating one possible approach to doing this experiment.

First fill a glass of water to the brim, and place an inverted saucer on top of it, as shown in Figure 3.13(a). Turn the glass upside down, then add water to the saucer and allow it to settle for half an hour. Add vegetable dye to the saucer, and finally lift the glass a tiny fraction to allow the dye to seep under the lip of the glass. This step is shown in Figure 3.13(b). Over a period of several days, the dye will diffuse into the glass of water. This is shown in the sequence of photographs in Figure 3.13(c), the initial state being the left-most picture.

The photographs and experimental procedure were kindly supplied by Dr Graham Read.

Figure 3.13 An experiment demonstrating diffusion.

Solutions to Exercises in Chapter 3

Solution 3.1

(a) $\nabla^2 f = 2 + 2 + 2 = 6$

(b) $\nabla^2 f = 2 - 2 = 0$

(c) $f = 1/\sqrt{x^2 + y^2 + z^2} = (x^2 + y^2 + z^2)^{-1/2}$, so

$$\frac{\partial f}{\partial x} = -x(x^2 + y^2 + z^2)^{-3/2},$$

$$\frac{\partial^2 f}{\partial x^2} = -(x^2 + y^2 + z^2)^{-3/2} + 3x^2(x^2 + y^2 + z^2)^{-5/2} = \frac{2x^2 - y^2 - z^2}{(x^2 + y^2 + z^2)^{5/2}}.$$

Similarly,

$$\frac{\partial^2 f}{\partial y^2} = \frac{2y^2 - z^2 - x^2}{(x^2 + y^2 + z^2)^{5/2}}, \qquad \frac{\partial^2 f}{\partial z^2} = \frac{2z^2 - x^2 - y^2}{(x^2 + y^2 + z^2)^{5/2}}$$

so

$$\nabla^2 f = \frac{(2x^2 - y^2 - z^2) + (2y^2 - z^2 - x^2) + (2z^2 - x^2 - y^2)}{(x^2 + y^2 + z^2)^{5/2}} = 0.$$

Solution 3.2

(a) The diffusion equation in two dimensions is

$$\frac{\partial \rho}{\partial t} = D\left[\frac{\partial^2 \rho}{\partial x^2} + \frac{\partial^2 \rho}{\partial y^2}\right].$$

(b) Consider the trial solution $\rho(x, y, t) = A(t) \exp[-\beta(x^2 + y^2)/t]$. We find

$$\frac{\partial \rho}{\partial t} = A'(t) \exp[-\beta(x^2 + y^2)/t] + \frac{\beta(x^2 + y^2)}{t^2} A(t) \exp[-\beta(x^2 + y^2)/t],$$

$$\frac{\partial \rho}{\partial x} = A(t) \frac{-2\beta x}{t} \exp[-\beta(x^2 + y^2)/t],$$

$$\frac{\partial^2 \rho}{\partial x^2} = \frac{-2\beta A(t)}{t} \exp[-\beta(x^2 + y^2)/t] + \frac{4\beta^2 x^2}{t^2} A(t) \exp[-\beta(x^2 + y^2)/t].$$

The expression for $\partial^2 \rho/\partial y^2$ is obtained by exchanging x and y in the expression for $\partial^2 \rho/\partial x^2$. Substituting into the diffusion equation and cancelling the factor $\exp[-\beta(x^2 + y^2)/t]$ gives

$$A'(t) + \frac{\beta(x^2 + y^2)}{t^2} A(t) = D\left[\frac{-4\beta A(t)}{t} + \frac{4\beta^2}{t^2}(x^2 + y^2) A(t)\right].$$

This equation must be satisfied for all x, y and t. For any given value of t, the terms which are independent of x and y on either side of this equation must be equal. Similarly, the terms proportional to $(x^2 + y^2)/t^2$ must be equal, giving $\beta A(t) = 4D\beta^2 A(t)$, so $\beta = 1/4D$. Equating the terms that are independent of x and y gives

$$A'(t) = -\frac{4\beta D}{t} A(t) = -\frac{A(t)}{t},$$

which is a differential equation to be solved for $A(t)$. It may be integrated to give $\ln A = -\ln t + \text{constant}$ (for $A > 0$ and $t > 0$). Exponentiating gives $A = C/t$ for some arbitrary constant C. The diffusion equation therefore has a solution of the form

$$\rho(x, y, t) = \frac{C}{t} \exp[-(x^2 + y^2)/4Dt].$$

Solution 3.3

In one dimension the concentration must be defined as the number of particles per unit length.

For a uniform distribution of N particles over an interval of length L, the concentration is $c = N/L$.

When the distribution of particles is non-uniform, the concentration $c(x,t)$ at position x and time t can be defined in terms of the number of particles $\Delta N(x, \Delta x, t)$ in a short interval between $x - \tfrac{1}{2}\Delta x$ and $x + \tfrac{1}{2}\Delta x$, divided by the length of this interval:
$$c(x,t) = \frac{\Delta N(x, \Delta x, t)}{\Delta x}.$$
If the quantity $c(x,t)$ is almost independent of Δx for a wide range of values of Δx, the concentration is well-defined.

Similarly, in two dimensions the concentration is the number of particles per unit area.

If we interpret people as 'particles', population densities quoted in geographical atlases are concentrations of people. Thus, if there are ΔN people living inside a square region of side Δx, the population density is $c = \Delta N/((\Delta x)^2)$. The length Δx must be chosen appropriately; for example, when estimating the population density in an urban area, it might be appropriate to take $\Delta x = 1\,\mathrm{km}$, but not $\Delta x = 10\,\mathrm{m}$.

Solution 3.4

Using the one-dimensional version of equation (3.7), we have
$$\begin{aligned} N(a) &= \int_0^a dx \left[c_0 + c_1 \cos\left(\frac{\pi x}{L}\right) \exp(-\alpha t) \right] \\ &= c_0 a + \frac{c_1 L}{\pi} \exp(-\alpha t) \left[\sin\left(\frac{\pi x}{L}\right) \right]_0^a \\ &= c_0 a + \frac{c_1 L \sin(\pi a/L)}{\pi} \exp(-\alpha t). \end{aligned}$$
When $a = L$, we find $N(L) = c_0 L$, which is independent of time.

Solution 3.5

The volume is $500\,\mathrm{ml} = 0.5\,\mathrm{l} = 0.5 \times 10^{-3}\,\mathrm{m}^3$. The concentration is therefore
$$\begin{aligned} c &= \frac{10^{20}}{0.5 \times 10^{-3}} \text{ molecules per m}^3 \\ &= 2 \times 10^{23} \text{ molecules per m}^3 \\ &= 2 \times 10^{20} \text{ molecules per litre} \\ &= \frac{2 \times 10^{20}}{6.022 \times 10^{23}} \text{ mols per litre} \\ &\simeq 3.32 \times 10^{-4} \text{ mol l}^{-1}. \end{aligned}$$
The concentration can be expressed as either $2 \times 10^{23}\,\mathrm{m}^{-3}$ or $3.32 \times 10^{-4}\,\mathrm{mol\,l}^{-1}$.

Solution 3.6

Using the fact that $1\,\mathrm{m}^3 = 10^3\,\mathrm{l}$, we have
$$250\,\mu\mathrm{g\,l}^{-1} = 250 \times 10^{-6}\,\mathrm{g\,l}^{-1} = 250 \times 10^{-9}\,\mathrm{kg\,l}^{-1} = 250 \times 10^{-6}\,\mathrm{kg\,m}^{-3}.$$
The mass of a carbon dioxide molecule is 7.31×10^{-26} kg, so
$$\text{number of molecules per m}^3 \text{ of carbon dioxide} = \frac{250 \times 10^{-6}}{7.31 \times 10^{-26}} \simeq 3.42 \times 10^{21}.$$
Similarly,
$$\text{number of molecules per m}^3 \text{ of carbon monoxide} = \frac{2 \times 10^{-6}}{4.65 \times 10^{-26}} \approx 4.30 \times 10^{19}.$$

Solution 3.7

The flux of fish is 600 per day. To express this in terms of the SI unit of time, we note that one day is $60 \times 60 \times 24 = 86\,400$ seconds. In the appropriate SI unit, the flux is $\Phi = 600/86\,400 \simeq 0.0069\,\text{s}^{-1}$.

The cross-sectional area of the river is $A = 5 \times 1 = 5\,\text{m}^2$, so the flux density is $J = \Phi/A \simeq 0.0069/5\,\text{m}^{-2}\,\text{s}^{-1} \simeq 0.0014\,\text{m}^{-2}\,\text{s}^{-1}$.

Solution 3.8

The approximate number of hailstones collected is $n = 100/0.05 = 2000$. The radius of the pan is $r = 0.1\,\text{m}$, so the area is $A = \pi r^2 = \pi \times (0.1)^2 \simeq 0.0314\,\text{m}^2$. The time is $t = 5 \times 60 = 300\,\text{s}$. So the flux density is

$$J = \frac{n}{At} = \frac{2000}{\pi \times (0.1)^2 \times 300} \simeq 212\,\text{m}^{-2}\,\text{s}^{-1}\,.$$

Solution 3.9

(a) Using equation (3.20), the number of particles is

$$\begin{aligned}
N &= \int_V dV\; c(\mathbf{r},t) \\
&= \int_0^1 dx \int_0^1 dy \int_0^1 dz\; \left[1 + \tfrac{1}{10}x^2 + \tfrac{1}{4}xy^2\right] \\
&= \int_0^1 dx \left(1 + \tfrac{1}{10}x^2 + \tfrac{1}{4}x \left[\tfrac{1}{3}y^3\right]_0^1\right) \\
&= \int_0^1 dx\; \left[1 + \tfrac{1}{10}x^2 + \tfrac{1}{12}x\right] \\
&= \left[x + \tfrac{1}{30}x^3 + \tfrac{1}{24}x^2\right]_0^1 \\
&= \frac{129}{120}.
\end{aligned}$$

This result is not a large number, so if the concentration is defined as the number of particles per unit volume, it is meaningless. However, if the concentration is measured in mols per unit volume, this figure represents approximately $6.02 \times 10^{23} \times 129/120$ particles, which is a large number. In this case the number of particles within the cube is likely to be similar in magnitude to the calculated result, so the answer is meaningful.

(b) The flux is calculated using equation (3.21). Let $\mathbf{J} = J_1\mathbf{i} + J_2\mathbf{j} + J_3\mathbf{k}$. Because the normal to the surface is aligned with the unit vector $\mathbf{i}$, we have $\mathbf{J}\cdot d\mathbf{A} = J_1\,dy\,dz$. The integral is over the square region $x = 1$, $0 < y < 1$, $0 < z < 1$, so the flux is

$$\begin{aligned}
\Phi &= \int_S d\mathbf{A}\cdot\mathbf{J}(\mathbf{r},t) \\
&= \int_0^1 dy \int_0^1 dz\; \mathbf{i}\cdot\mathbf{J}(1,y,z,t) \\
&= \int_0^1 dy \int_0^1 dz\; J_1(1,y,z,t) \\
&= \int_0^1 dy \int_0^1 dz\; \left(2 + \tfrac{1}{10}z + \tfrac{1}{8}\right) \\
&= \int_0^1 dy \int_0^1 dz\; \left(\tfrac{17}{8} + \tfrac{1}{10}z\right) \\
&= \left[\tfrac{17}{8}z + \tfrac{1}{20}z^2\right]_0^1 \\
&= \frac{87}{40}.
\end{aligned}$$

Solution 3.10

The volume of the region bounded between two concentric spheres of radii r and $r + \delta r$ is

$$\delta V = \tfrac{4}{3}\pi(r+\delta r)^3 - \tfrac{4}{3}\pi r^3 = 4\pi r^2 \delta r + O((\delta r)^2).$$

The contribution to the number of particles from this spherical shell is $c(r)\,\delta V$. Integrating over r, the number of particles inside a sphere of radius R is

$$N(R) = 4\pi \int_0^R dr\, r^2 c(r).$$

When $c(r) = 1 + 1/r$, this is

$$N(R) = 4\pi \int_0^R dr\, r^2 \left(1 + \frac{1}{r}\right) = 4\pi\left[\tfrac{1}{3}r^3 + \tfrac{1}{2}r^2\right]_0^R = \tfrac{4}{3}\pi R^3 + 2\pi R^2.$$

This result could also have been obtained by working in spherical polar coordinates and integrating over the angle variables.

Solution 3.11

The area of a spherical surface of radius $r = R$ is $4\pi R^2$. The flux density vector is always aligned with the outward normal, and has uniform magnitude $J(R)$ across this surface. The flux across this spherical surface is therefore

$$\Phi(R) = 4\pi R^2 J(R).$$

Clearly, if $J = C/R^2$ for some constant C, then the flux $\Phi = 4\pi C$ is independent of R.

Solution 3.12

If $\boldsymbol{F} = F_1 \mathbf{i} + F_2 \mathbf{j} + F_3 \mathbf{k}$, then

$$\begin{aligned}
\boldsymbol{\nabla} \cdot (a\boldsymbol{F}) &= \frac{\partial(aF_1)}{\partial x} + \frac{\partial(aF_2)}{\partial y} + \frac{\partial(aF_3)}{\partial z} \\
&= a\frac{\partial F_1}{\partial x} + F_1\frac{\partial a}{\partial x} + a\frac{\partial F_2}{\partial y} + F_2\frac{\partial a}{\partial y} + a\frac{\partial F_3}{\partial z} + F_3\frac{\partial a}{\partial z} \\
&= a\boldsymbol{\nabla}\cdot\boldsymbol{F} + \boldsymbol{F}\cdot\boldsymbol{\nabla}a.
\end{aligned}$$

Solution 3.13

The continuity equation for the liquid is

$$\frac{\partial \rho_\mathrm{m}}{\partial t} + \boldsymbol{\nabla}\cdot(\rho_\mathrm{m}\boldsymbol{v}) = 0.$$

Since ρ_m is constant, we have $\partial \rho_\mathrm{m}/\partial t = 0$ and $\boldsymbol{\nabla}\rho_\mathrm{m} = 0$. Using the result of the previous exercise,

$$\boldsymbol{\nabla}\cdot(\rho_\mathrm{m}\boldsymbol{v}) = \rho_\mathrm{m}\boldsymbol{\nabla}\cdot\boldsymbol{v} + \boldsymbol{v}\cdot\boldsymbol{\nabla}\rho_\mathrm{m}.$$

The second term in this equation is zero because ρ_m is constant. The continuity equation then reduces to

$$\boldsymbol{\nabla}\cdot\boldsymbol{v} = 0.$$

Solution 3.14

Because the flow is steady, $\partial c/\partial t = 0$, and the continuity equation becomes $\nabla \cdot \boldsymbol{J} = 0$.

(a) $\nabla \cdot \boldsymbol{J} = 1 + 1 + 1 = 3$. This vector field cannot be the flux density in a steady flow.

(b) $\nabla \cdot \boldsymbol{J} = 0 + 0 + 0 = 0$. This vector field can be the flux density of a steady flow.

(c) $\nabla \cdot \boldsymbol{J} = A + D$. This vector field can be the flux density of a steady flow only if $A + D = 0$.

Solution 3.15

(a) For $x = 0$, the outward normal is $\boldsymbol{n} = -\mathbf{i}$. The vector field at $x = 0$ is $\boldsymbol{F} = y^2 \mathbf{i} + 2y^2 \mathbf{j}$, so $\boldsymbol{F} \cdot \boldsymbol{n} = -y^2$. The surface integral is

$$\int_{x=0} d\boldsymbol{A} \cdot \boldsymbol{F} = \int_0^1 dy \int_0^1 dz\, \boldsymbol{F} \cdot \boldsymbol{n} = \int_0^1 dy \int_0^1 dz\, (-y^2) = -\int_0^1 dy\, y^2 = -\tfrac{1}{3}.$$

For $x = 1$, the outward normal is $\boldsymbol{n} = \mathbf{i}$, and $\boldsymbol{F} = (1 + y^2)\mathbf{i} + (2y^2 - 1)\mathbf{j}$, so $\boldsymbol{F} \cdot \boldsymbol{n} = 1 + y^2$ and

$$\int_{x=1} d\boldsymbol{A} \cdot \boldsymbol{F} = \int_0^1 dy \int_0^1 dz\, (1 + y^2) = \int_0^1 dy\, (1 + y^2) = \tfrac{4}{3}.$$

For $y = 0$, the outward normal is $\boldsymbol{n} = -\mathbf{j}$, and $\boldsymbol{F} = x\mathbf{i} - x^3\mathbf{j}$, so $\boldsymbol{F} \cdot \boldsymbol{n} = x^3$ and

$$\int_{y=0} d\boldsymbol{A} \cdot \boldsymbol{F} = \int_0^1 dx \int_0^1 dz\, x^3 = \tfrac{1}{4}.$$

For $y = 1$, the outward normal is $\boldsymbol{n} = \mathbf{j}$, and $\boldsymbol{F} = (1 + x)\mathbf{i} + (2 - x^3)\mathbf{j}$, so $\boldsymbol{F} \cdot \boldsymbol{n} = 2 - x^3$ and

$$\int_{y=1} d\boldsymbol{A} \cdot \boldsymbol{F} = \int_0^1 dx \int_0^1 dz\, (2 - x^3) = \tfrac{7}{4}.$$

For $z = 0$, the outward normal is $\boldsymbol{n} = -\mathbf{k}$ and $\boldsymbol{F} \cdot \boldsymbol{n} = 0$. The integral over the surface $z = 0$ is therefore zero. Similarly, the integral over $z = 1$ vanishes.

The integral over the surface S of the cube is therefore

$$\Phi_S = \int_S d\boldsymbol{A} \cdot \boldsymbol{F} = -\tfrac{1}{3} + \tfrac{4}{3} + \tfrac{1}{4} + \tfrac{7}{4} + 0 + 0 = 3.$$

(b) We have

$$\nabla \cdot \boldsymbol{F} = \frac{\partial}{\partial x}(x + y^2) + \frac{\partial}{\partial y}(2y^2 - x^3) + \frac{\partial}{\partial z}(0) = 1 + 4y.$$

The volume integral of this divergence is

$$\Phi_V = \int_V dV\, \nabla \cdot \boldsymbol{F}$$

$$= \int_0^1 dx \int_0^1 dy \int_0^1 dz\, (1 + 4y)$$

$$= \int_0^1 dx \int_0^1 dy\, (1 + 4y) = \int_0^1 dy\, (1 + 4y) = \left[y + 2y^2\right]_0^1 = 3.$$

Thus $\Phi_V = \Phi_S$, and Gauss's theorem is satisfied.

Solution 3.16

The number of particles is

$$N = \int_0^\infty dx\, c(x,t) = \frac{1}{t}\int_0^\infty dx\, \exp(-x/t) = 1.$$

(Because this is not a large number, we must assume that N is measured in multiples of some large number, such as the mol.) Hence N is constant, so the number of particles is conserved.

Because the number of particles is conserved (that is, N is independent of time), we are justified in using the continuity equation: we find

$$\frac{\partial J(x,t)}{\partial x} = -\frac{\partial c(x,t)}{\partial t} = \frac{1}{t^3}(t-x)\exp(-x/t).$$

We have $J(0,t) = 0$, because the particles cannot cross $x = 0$. Integrating gives

$$J(x,t) = \int_0^x du\, \frac{1}{t^3}(t-u)\exp(-u/t)$$

where the lower limit of integration ensures that the condition $J(0,t) = 0$ is satisfied. Using integration by parts, this gives

$$J(x,t) = \left[(t-y)\left(-\frac{1}{t^2}\right)\exp(-y/t)\right]_0^x - \frac{1}{t^2}\int_0^x dy\, \exp(-y/t)$$

$$= \frac{x-t}{t^2}\exp(-x/t) + \frac{1}{t} + \frac{1}{t}[\exp(-y/t)]_0^x$$

$$= \frac{x}{t^2}\exp(-x/t).$$

This is clearly not proportional to $\partial c/\partial x = -\exp(-x/t)/t^2$, so Fick's law is not satisfied for any choice of diffusion constant D.

Given that $c(x,t)$ is a concentration, it must represent the concentration of a material which is not moving by a process of diffusion.

Solution 3.17

From the definition of the gradient vector discussed in Block 0, Chapter 2, the gradient of the function $f(r)$ is $f'(r)\hat{\mathbf{r}}$, where $f'(r)$ is the derivative of f and $\hat{\mathbf{r}}$ is a unit vector pointing radially outwards. If $f(r) = 1/r$, the gradient is

$$\boldsymbol{\nabla} f = -\frac{1}{r^2}\hat{\mathbf{r}}.$$

You have seen in exercise 3.11 that the flux of this vector field across a spherical surface of radius R centred on the origin is independent of R. The surface integral of $\boldsymbol{\nabla} f$ over a spherical surface of radius R_1 is therefore the same as that over a surface of radius R_2. Applying Gauss's theorem, the difference between these two surface integrals is the volume integral of the divergence of the vector field $\boldsymbol{\nabla} f$ over the region between radii R_1 and R_2. (The difference enters here because $d\mathbf{A}$ points in the direction of the outward normal to the bounding surface, so it points away from the origin for the outer sphere but towards the origin for the inner sphere.) The integral of the divergence of the vector field, i.e. of $\nabla^2 f$, over this region is zero. Because of the spherical symmetry, $\nabla^2 f$ can depend only upon the distance from the origin. Allowing R_2 to approach R_1, and noting that $\nabla^2 f$ must be a continuous function of r, we see that $\nabla^2 f = 0$ at R_1 in order for the integral to equal zero. We therefore must have $\nabla^2 f = 0$ for all points on the spherical surface at R_1. Because R_1 is arbitrary, this argument establishes that $\nabla^2 f = 0$ everywhere except at the origin, $r = 0$.

Solution 3.18

(a) Because of the spherical symmetry of the problem, the flux density on a spherical surface at R takes the same value $J(R,t)$ everywhere on that surface. The flux $\Phi(R)$ across this surface is then the flux density multiplied by the area of this surface: $\Phi(R) = 4\pi R^2 J(R,t)$. Because the total number of particles is conserved, the outward flux across this surface is equal to the decrease in the number of particles within this surface, which is obtained by differentiating equation (3.7) with respect to time. Thus, we obtain

$$\Phi(R) = 4\pi R^2 J(R,t) = -\frac{\partial}{\partial t}\left[\int_0^R dr\, 4\pi r^2 c(r,t)\right],$$

since

$$N(R,t) = \int_0^R dr\, 4\pi r^2 c(r,t)$$

is the total number of particles, $N(R,t)$, inside the sphere. The flux density is therefore obtained from the concentration via

$$J(r,t) = -\frac{1}{r^2}\int_0^r ds\, s^2 \frac{\partial c}{\partial t}(s,t).$$

after a change in variable names (R to r and r to s inside the integral).

(b) If the concentration satisfies the diffusion equation, the flux density J must (according to Fick's law) be proportional to the gradient of the concentration. The magnitude of the gradient of the concentration in the radial direction is just $\partial c/\partial r$. We therefore have

$$-D\frac{\partial c}{\partial r}(r,t) = J = -\frac{1}{r^2}\int_0^r ds\, s^2 \frac{\partial c}{\partial t}(s,t).$$

Multiplying both sides by $-r^2$, then taking the derivative of both sides with respect to r, and finally dividing both sides by r^2, gives the result quoted in the exercise.

Solution 3.19

(a) The SI unit of energy is the Joule, where $1\,\text{J} = 1\,\text{kg}\,\text{m}^2\,\text{s}^{-2}$, so the SI unit of energy per unit volume q is therefore $\text{J}\,\text{m}^{-3}$.

(b) From equation (3.52), the SI unit of mass-specific heat capacity C is the unit of $(q - q_0)/\rho_m(\theta - \theta_0)$. Since the SI unit of temperature is the Kelvin, and those of q and ρ_m are $\text{J}\,\text{m}^{-3}$ and $\text{kg}\,\text{m}^{-3}$ respectively, the SI unit for C is $\text{J}\,\text{m}^{-3}/\text{kg}\,\text{m}^{-3}\,\text{K} = \text{J}\,\text{kg}^{-1}\,\text{K}^{-1}$.

Solution 3.20

(a) The thermal diffusivity of air is

$$D = \frac{\kappa}{C\rho_m} = \frac{2.4 \times 10^{-2}}{1.29 \times 1.01 \times 10^3} \simeq 1.84 \times 10^{-5}\,\text{m}^2\,\text{s}^{-1}.$$

(b) The thermal diffusivity D calculated in part (a) is similar in magnitude to the diffusion constants for gases diffusing through air quoted in the exercise. The similarity of these values indicates that the diffusion of heat energy has the same origin as the diffusion of materials: heat is transported by the random microscopic motion of atoms.

Solution 3.21

If
$$c = \frac{N}{\sqrt{4\pi Dt}} \exp(-x^2/4Dt),$$

then
$$\frac{\partial c}{\partial t} = -\frac{N}{2\sqrt{4\pi D}\, t^{3/2}} \exp(-x^2/4Dt) + \frac{x^2}{4Dt^2} \frac{N}{\sqrt{4\pi Dt}} \exp(-x^2/4Dt),$$
$$\frac{\partial c}{\partial x} = \frac{-xN}{2Dt\sqrt{4\pi Dt}} \exp(-x^2/4Dt),$$
$$\frac{\partial^2 c}{\partial x^2} = -\frac{N}{2Dt\sqrt{4\pi Dt}} \exp(-x^2/4Dt) + \frac{x^2 N}{4D^2 t^2 \sqrt{4\pi Dt}} \exp(-x^2/4Dt)$$
$$= \frac{1}{D}\frac{\partial c}{\partial t}$$

so that equation (3.1) is satisfied.

Solution 3.22

Consider the function $c_1(x,t) = c(x-x_0, t-t_0)$. We have
$$\frac{\partial c_1}{\partial t}(x,t) = \frac{\partial c}{\partial t}(x-x_0, t-t_0),$$

and analogous relations hold for the first and second derivatives with respect to x. If c satisfies the diffusion equation at position $x - x_0$ and time $t - t_0$, then c_1 satisfies the diffusion equation at position x and time t. It was assumed that c satisfies the diffusion equation for all positions and times. It follows that $c_1(x,t)$ also satisfies the diffusion equation for all positions and times.

Solution 3.23

To check normalisation, we calculate
$$\int_{-\infty}^{\infty} dx\, \rho(x,t) = \frac{1}{\sqrt{4\pi Dt}} \int_{-\infty}^{\infty} dx\, \exp(-x^2/4Dt)$$
$$= \frac{1}{\sqrt{\pi}} \int_{-\infty}^{\infty} du\, \exp(-u^2) = 1.$$

We used the change of variable $u = x/\sqrt{4Dt}$, and applied the standard Gaussian integral, discussed in Block 0, Chapters 1 and 2.

The second moment is given by
$$\int_{-\infty}^{\infty} dx\, x^2 \rho(x,t) = \frac{1}{\sqrt{4\pi Dt}} \int_{-\infty}^{\infty} dx\, x^2 \exp(-x^2/4Dt)$$
$$= \frac{4Dt}{\sqrt{\pi}} \int_{-\infty}^{\infty} du\, u^2 \exp(-u^2) = \frac{4Dt}{\sqrt{\pi}} \frac{\sqrt{\pi}}{2} = 2Dt.$$

Again, the change of variable is $u = x/\sqrt{4Dt}$, and we used the integral I_2 defined in Section 1.3.

Solution 3.24

(a) Given two functions $f(x,t)$ and $g(x,t)$, both of which are solutions of the one-dimensional diffusion equation, the function
$$c(x,y,t) = f(x,t)\, g(y,t)$$

is a solution of the two-dimensional diffusion equation. This is verified as follows:
$$\frac{\partial c}{\partial t} = \frac{\partial f(x,t)}{\partial t} g(y,t) + f(x,t) \frac{\partial g(y,t)}{\partial t}$$
$$= D\frac{\partial^2 f(x,t)}{\partial x^2} g(y,t) + D f(x,t) \frac{\partial^2 g(y,t)}{\partial y^2}$$
$$= D\frac{\partial^2 c}{\partial x^2} + D\frac{\partial^2 c}{\partial y^2}.$$

The first step uses the product rule for differentiation; the second uses the fact that both f and g satisfy the one-dimensional diffusion equation.

(b) Consider the case where $g = f$ in the notation of part (a), so $\rho(x,y,t) = f(x,t)f(y,t)$. If $f(x,t)$ is a normalised solution of the diffusion equation in one dimension, then

$$\int_{-\infty}^{\infty} dx \int_{-\infty}^{\infty} dy\, \rho(x,y,t) = \left[\int_{-\infty}^{\infty} dx\, f(x,t)\right] \times \left[\int_{-\infty}^{\infty} dy\, f(y,t)\right] = 1 \times 1,$$

so the two-dimensional solution is automatically normalised. From equation (3.60) and Exercise 3.23, we have a normalised one-dimensional solution of the form $f(x,t) = \exp(-x^2/4Dt)/\sqrt{4\pi Dt}$, and this leads to the two-dimensional solution quoted in the exercise with the constant C given by $1/4\pi D$.

(c) The mean value of $r^2 = x^2 + y^2$ is

$$\int_{-\infty}^{\infty} dx \int_{-\infty}^{\infty} dy\, (x^2 + y^2)\rho(x,y,t) = \int_{-\infty}^{\infty} dx\, x^2 f(x,t) \int_{-\infty}^{\infty} dy\, f(y,t)$$
$$+ \int_{-\infty}^{\infty} dx\, f(x,t) \int_{-\infty}^{\infty} dy\, y^2 f(y,t)$$
$$= 2Dt \times 1 + 1 \times 2Dt$$
$$= 4Dt.$$

The final step uses the fact that the second moment of the function $f(x,t)$ used in part (b) is $2Dt$ (see Exercise 3.23).

(d) In three dimensions, the same approach works. It turns out that $\langle r^2 \rangle = 6Dt$ in three dimensions (as might be expected by extrapolation from the one- and two-dimensional cases).

Solution 3.25

The flux densities in regions 1 and 2 are respectively $\boldsymbol{J}_1 = -D_1 \boldsymbol{\nabla} c_1$ and $\boldsymbol{J}_2 = -D_2 \boldsymbol{\nabla} c_2$, where c_1 and c_2 are the concentrations in each region. Let $\boldsymbol{n}$ be the normal to the interface, pointing from region 1 to region 2. Because particles are neither created nor destroyed at the interface, the flux density of particles crossing the interface on either side must be the same. The flux density of particles striking the interface from region 1 is $J_1 = \boldsymbol{J}_1 \cdot \boldsymbol{n}$. The flux density J_2 in region 2 is obtained in the same way. Equating these gives $\boldsymbol{J}_1 \cdot \boldsymbol{n} = \boldsymbol{J}_2 \cdot \boldsymbol{n}$, so

$$(D_1 \boldsymbol{\nabla} c_1 - D_2 \boldsymbol{\nabla} c_2) \cdot \boldsymbol{n} = 0.$$

This condition holds for all points on the interface, and for all times.

Solution 3.26

We assume one-dimensional flow in the x direction (chosen to be perpendicular to the boundaries), so the temperature θ satisfies $\partial\theta/\partial t = D\, \partial^2 \theta/\partial x^2$, where $D = \kappa/C\rho_{\mathrm{m}}$. The origin is taken to be at the interface between the layers. Because the flow is steady, $\partial\theta/\partial t = 0$, so $\partial^2 \theta/\partial x^2 = 0$. The temperature in layer j is thus $\theta_j(x) = A_j x + B_j$, for some constants A_j and B_j which may take different values in different layers. The flux density of heat is $J = -\kappa_j \partial\theta_j/\partial x = -\kappa_j A_j$, which is the same in each layer. The temperature difference across layer j is $\Delta\theta_j = A_j L_j = -JL_j/\kappa_j$. (Here, $\Delta\theta_1 = \theta_{\mathrm{i}} - \theta_1$ and $\Delta\theta_2 = \theta_2 - \theta_{\mathrm{i}}$.)

The overall temperature difference across both layers is $\Delta\theta_1 + \Delta\theta_2$. We are told that this difference is $\theta_2 - \theta_1$, so

$$\theta_2 - \theta_1 = -J\left[\frac{L_1}{\kappa_1} + \frac{L_2}{\kappa_2}\right] = -J\frac{L_1\kappa_2 + L_2\kappa_1}{\kappa_1\kappa_2},$$

and the heat flux density is thus

$$J = \frac{(\theta_1 - \theta_2)\kappa_1\kappa_2}{L_1\kappa_2 + L_2\kappa_1}.$$

The temperature difference across the first layer is then $\Delta\theta_1 = -JL_1/\kappa_1$. The interface temperature is $\theta_i = \theta_1 + \Delta\theta_1$, i.e.

$$\theta_i = \theta_1 + \frac{(\theta_2 - \theta_1)L_1\kappa_2}{L_1\kappa_2 + L_2\kappa_1}.$$

Solution 3.27

(a) As discussed in the question (see the Hint), the heat flux is the same across all cylindrical surfaces. The outward heat flux Φ across a surface of area A due to a temperature gradient $d\theta/dr$ is

$$\Phi = -\kappa A \frac{d\theta}{dr}.$$

In this case the area of the curved part of a cylindrical surface is $2\pi r L$, where L is the length of the cylinder, and r is its radius. The heat flux is then

$$\Phi = -2\pi\kappa r L \frac{d\theta}{dr} = \text{constant}.$$

This is a differential equation for $\theta(r)$, which gives

$$\int d\theta = -(\Phi/2\pi\kappa L) \int dr/r.$$

Performing the integrations, we obtain

$$\theta(r) = -\frac{\Phi}{2\pi\kappa L} \ln r + \theta_0,$$

for some constant θ_0. This solution contains two unknown constants, θ_0 and Φ, which are determined by applying the boundary conditions that the temperatures are θ_1, θ_2 at radii r_1 and r_2, respectively.

Writing $\theta_0 = \theta_1 + (\Phi/2\pi\kappa L) \ln r_1$, the solution reads

$$\theta(r) = \theta_1 - \frac{\Phi}{2\pi\kappa L} \ln\left(\frac{r}{r_1}\right).$$

This is seen to satisfy the condition $\theta(r_1) = \theta_1$. If the heat flux Φ is now chosen so that

$$\frac{\Phi}{2\pi\kappa L} = \frac{\theta_1 - \theta_2}{\ln(r_2/r_1)},$$

the condition $\theta(r_2) = \theta_2$ is also satisfied, and $\theta(r)$ has the form quoted in the exercise.

(b) The heat flux per unit length of pipe is

$$\frac{\Phi}{L} = \frac{2\pi\kappa(\theta_1 - \theta_2)}{\ln(r_2/r_1)}.$$

Because the logarithmic function increases very slowly as a function of r_2/r_1, this result shows that it is difficult to reduce heat loss significantly by increasing the thickness of insulation. Doubling the thickness of insulation does not halve the heat loss (except in the limit as $r_2/r_1 \to 1$).

Solution 3.28

In time δt, the volume of fluid passing through an element of area $\Delta\mathbf{A}$ with velocity $\mathbf{v}$ is $\delta V = \mathbf{v} \cdot \Delta\mathbf{A}\,\delta t$. If the concentration of particles is c, the number of particles in this volume is $\delta N = c\,\delta V$. But, from the definition of flux density, $\delta N = \mathbf{J}_{\text{flow}} \cdot \Delta\mathbf{A}\,\delta t$. Since this should hold for any choice of $\Delta\mathbf{A}$, it follows that $\mathbf{J}_{\text{flow}} = c\mathbf{v}$.

This so-called advective contribution to the flux density must be added to the diffusive flux density given by Fick's law. The total flux density is therefore $\mathbf{J} = \mathbf{v}c - D\,\nabla c$. Applying the continuity equation and using the result of Exercise 3.12 gives

$$\frac{\partial c}{\partial t} = -\nabla \cdot \left[\mathbf{v}c - D\,\nabla c\right] = D\,\nabla \cdot \nabla c - \mathbf{v} \cdot \nabla c - (\nabla \cdot \mathbf{v})c,$$

which is equivalent to the equation quoted in the exercise.

Solution 3.29

(a) For a steady-state solution, the concentration c is a function of position x only. The diffusion–advection equation in one dimension may then be written

$$0 = \frac{d}{dx}\left[v(x)c(x) - D\frac{dc}{dx}\right],$$

so $v(x)c(x) - Dc'(x) = C$, where C is a constant. If c approaches zero as $x \to \infty$, we expect that its derivative $c'(x)$ also approaches zero as $x \to \infty$. This indicates that we must have $C = 0$. So the differential equation to be solved is

$$\frac{dc}{dx} = \frac{v(x)c(x)}{D},$$

which can be solved by, for example, separation of variables:

$$\int \frac{dc}{c} = \int dx\, \frac{v(x)}{D},$$

giving

$$\ln c(x) + \text{constant} = \int_0^x dy\, \frac{v(y)}{D}.$$

Exponentiating this last equation gives the required solution:

$$c(x) = A \exp\left[\frac{1}{D}\int_0^x dy\, v(y)\right],$$

for some constant A.

(b) Using equation (3.7), N and A are related by

$$N = \int_0^\infty dx\, c(x) = A \int_0^\infty dx\, \exp\left[\frac{1}{D}\int_0^x dy\, v(y)\right].$$

Thus A can be determined from an integral.

(c) (i) For $v(x) = -\lambda$, since

$$\int_0^x dy\, v(y) = -\lambda \int_0^x dy = -\lambda x,$$

we have

$$c(x) = A\exp(-\lambda x/D).$$

(ii) For $v(x) = -\mu x$, since

$$\int_0^x dy\, v(y) = -\mu \int_0^x dy\, y = -\tfrac{1}{2}\mu x^2,$$

we have

$$c(x) = A\exp(-\mu x^2/2D).$$

CHAPTER 4
Solutions of the diffusion equation

4.1 Introduction

In this chapter we survey many of the methods used to find solutions of the diffusion equation,

$$\frac{\partial c}{\partial t} = D\nabla^2 c. \tag{4.1}$$

We look at solutions for various types of boundary and initial conditions. The physical context in which each of the solutions arises is described. In some cases, it is more natural to think of the function $c(\boldsymbol{r}, t)$ as representing a temperature rather than a concentration, and in these cases we replace the symbol c by θ.

We show in detail how to construct solutions of the diffusion equation satisfying appropriate boundary conditions, but we shall not discuss why these solutions are unique. However, if the mathematical representation of a physical problem is correctly formulated, the solution is expected to be unique, so the question of uniqueness need not concern us. Also, some solutions are obtained as infinite series, but the convergence of these series is not discussed in any detail.

The diffusion equation and the wave equation are both linear partial differential equations, and many of the techniques of solution are common to both equations. Separation of variables and Fourier methods are often very useful in both cases; other techniques are also used, and many of the methods may be familiar from Block I. This chapter is, however, self-contained, apart from making frequent reference to Chapter 3 of Block I, which covers Fourier series and Fourier transforms. You may find it useful to briefly review that chapter before continuing; the sections on the convolution theorem (Section 3.5) and on the Fourier transform of derivatives (Section 3.6) are particularly relevant.

The discussion is organised according to the nature of the boundaries and the number of space dimensions of the 'medium' (the material body) within which the diffusion equation is satisfied. We consider successively an infinite medium, a finite medium and a semi-infinite medium. The first two of these are self-explanatory; a semi-infinite medium is obtained by dividing an infinite space into two infinite regions by a boundary, then taking the region on one side of this boundary. The only case of a semi-infinite medium that we consider is where the boundary is a flat surface (which we take to be the surface $x = 0$ in Cartesian coordinates), and we shall consider the solution of the diffusion equation in the half-space $x > 0$. The semi-infinite medium can be used to model phenomena close to the surface of a large medium, such as temperature variations in the soil close to the surface of the Earth.

4.2 Diffusion in an infinite medium

An *infinite medium* is a region of infinite extent, without boundaries, where a particular equation is satisfied. We have already encountered a solution of the diffusion equation in an infinite medium in one dimension, in Chapter 3 (Section 3.7). More generally, if all of the concentration is initially at x_0 at time $t = 0$, the solution of the diffusion equation for $t > 0$ is

$$c(x,t) = \frac{N}{\sqrt{4\pi Dt}} \exp\left[-\frac{(x-x_0)^2}{4Dt}\right], \qquad (4.2)$$

where N is the total number of particles. This solution is verified by checking that it solves the differential equation, and has the correct form at $t = 0$: as $t \to 0$ from above, the concentration approaches zero everywhere except at $x = x_0$. Examples of this solution (at different times and with $x_0 = 0$) were plotted in the previous chapter, as Figure 3.10. The solution should be considered undefined for $t < 0$. This solution could have been achieved by inspired guesswork, but it is instructive to see how it can be obtained by a systematic approach. This discussion will naturally lead us to a general solution of the diffusion equation in an infinite medium.

The function $c(x,t)$ may be expressed in terms of its Fourier transform with respect to x, i.e.

$$c(x,t) = \frac{1}{\sqrt{2\pi}} \int_{-\infty}^{\infty} dk\, \tilde{c}(k,t) \exp(ikx). \qquad (4.3)$$

This corresponds to the definition of the Fourier transform given as equation (3.24) in Block I, Chapter 3, with both functions depending on an additional variable, t. Now we make use of ideas developed in Section 3.6 of that chapter. If equation (4.3) is substituted into the one-dimensional form of the diffusion equation (4.1), and the derivatives are taken inside the integral sign, we conclude that $\tilde{c}(k,t)$ satisfies

$$\frac{\partial \tilde{c}}{\partial t}(k,t) = -Dk^2\, \tilde{c}(k,t). \qquad (4.4)$$

Exercise 4.1

Show that if $\tilde{c}(k,t)$ satisfies equation (4.4), then $c(x,t)$ satisfies the one-dimensional diffusion equation. [Hint: Note that $\partial^2/\partial x^2$ acts only upon the term $\exp(ikx)$, and that $\frac{\partial^2}{\partial x^2} \exp(ikx) = -k^2 \exp(ikx)$.]

The essential point is that the diffusion equation becomes easier to solve after taking the Fourier transform. In the Fourier-transformed representation, derivatives of $c(x,t)$ with respect to x are transformed into multiplication of $\tilde{c}(k,t)$ by powers of the Fourier conjugate variable, k. The Fourier transform of the concentration then satisfies an ordinary differential equation in t rather than a partial differential equation in x and t. (Equation (4.4) can be regarded as an ordinary differential equation because the *only* derivative present is that with respect to t, and k can be treated as if it were a constant.) This transformation is useful because ordinary differential equations are usually much easier to solve than partial differential equations.

4.2 Diffusion in an infinite medium

Exercise 4.2

Solve the equation

$$\frac{df}{dt} = -\alpha f$$

and use the solution to solve equation (4.4). Express your answer in terms of $\tilde{c}(k, 0)$, the Fourier transform of the initial concentration.

In Exercise 4.2, you should have obtained

$$\tilde{c}(k, t) = \exp(-Dk^2 t)\, \tilde{c}(k, 0) \tag{4.5}$$

where the function $\tilde{c}(k, 0)$ is the Fourier transform of the initial concentration, $c(x, 0)$ (we assume that this Fourier transform exists). This approach solves the problem of determining $c(x, t)$, provided that we can perform the two integrals required for the Fourier transform of $c(x, 0)$ and the inverse Fourier transform of $\tilde{c}(k, t)$. However, there is a more direct and more intuitive route to the solution, which we discover by applying the convolution theorem to equation (4.5).

According to equation (4.5), $\tilde{c}(k, t)$ may be expressed as a product of two known functions of k, namely $\exp(-Dk^2 t)$ and $\tilde{c}(k, 0)$. Recall the convolution theorem, considered in Block I, Chapter 3, Section 3.5, which gives an expression for the Fourier transform of a function $h = f \otimes g$ which is the convolution of two functions $f(x)$ and $g(x)$. This states that the Fourier transform of $h(x)$ is proportional to the product of the Fourier transforms of f and g: $\tilde{h}(k) = \sqrt{2\pi}\, \tilde{f}(k)\tilde{g}(k)$. We write equation (4.5) in the form

$$\tilde{c}(k, t) = \sqrt{2\pi}\, \tilde{K}(k, t)\, \tilde{c}(k, 0), \tag{4.6}$$

and apply the converse of the convolution theorem. We see that the concentration $c(x, t)$ must be the convolution of the initial concentration, $c(x, 0)$, and another function $K(x, t)$ which will be called the *propagator*:

$$c(x, t) = \int_{-\infty}^{\infty} dx_0\, K(x - x_0, t)\, c(x_0, 0)\ . \tag{4.7}$$

Comparison between equations (4.5) and (4.6) shows that the propagator $K(x, t)$ is the inverse Fourier transform of a Gaussian function: $\tilde{K}(k, t) = \exp(-Dk^2 t)/\sqrt{2\pi}$ (we consider only $t > 0$ throughout). Calculating the inverse Fourier transform, we find that $K(x, t)$ is also a Gaussian function:

$$K(x, t) = \frac{1}{2\pi} \int_{-\infty}^{\infty} dk\, \exp(ikx) \exp(-Dk^2 t)$$

$$= \frac{1}{\sqrt{4\pi Dt}} \exp(-x^2/4Dt). \tag{4.8}$$

(This Fourier transform pair can be deduced from Exercise 3.16 of Block I, Chapter 3, which shows that the Fourier transform of $\exp(-x^2/4Dt)$ is $\sqrt{2Dt}\exp(-Dtk^2)$. This implies that the Fourier transform of $K(x, t)$, as given by equation (4.8), is $\tilde{K}(k, t) = \exp(-Dtk^2)/\sqrt{2\pi}$, as required.) We also see that the function $K(x - x_0, t)$ corresponds to the special solution considered at the start of this section, equation (4.2), *divided* by the total number of particles, N. The propagator, $K(x - x_0, t)$, therefore has a simple interpretation: it is the *probability density function* that a particular particle reaches position x at time t, given that it started at position x_0 at time $t = 0$.

The propagator is a very useful concept for treating linear partial differential equations which arise in applied mathematics and theoretical physics. It determines how a 'field' (in this case, the concentration) at one point in space and time, (x_0, t_0), influences ('propagates' into) the solution at another point, (x, t). Knowledge of the propagator enables solutions to be found for any initial condition of the field. In the more general case where there are boundaries or where the coefficients of the linear partial differential equation depend upon position, the propagator is not a function of the difference between the final and initial positions, but depends upon x and x_0 separately. Similarly, when the boundaries move, the propagator can depend upon the initial and final times separately. In the case of a general linear partial differential equation in one dimension, equation (4.7) would be written in the form

$$c(x,t) = \int_{-\infty}^{\infty} dx_0 K(x, x_0, t, t_0) c(x_0, t_0). \tag{4.9}$$

Later we shall consider cases where the propagator depends upon x and x_0 separately, but in all of the examples considered in this course, the propagator depends only upon the time difference, $t - t_0$. (In the above, $t_0 = 0$.)

Example 4.1

The partial differential equation

$$\frac{\partial c}{\partial t} = D \frac{\partial^2 c}{\partial x^2} - Rc,$$

where D and R are constants, can be used to model the concentration of particles which disappear (for example, due to radioactive disintegration). It can also model heat flow along a rod, in cases where heat is lost into the surroundings. The constant R is called a *rate constant*.

Determine the propagator for this equation (taking the initial time, t_0, equal to zero).

Solution

Fourier transformation leads to an ordinary differential equation for $\tilde{c}(k,t)$:

$$\frac{\partial \tilde{c}}{\partial t} = (-Dk^2 - R)\tilde{c},$$

which has the solution

$$\tilde{c}(k,t) = \exp[-(Dk^2 + R)t]\,\tilde{c}(k,0),$$

so $\tilde{K}(k,t) = \exp[-(Dk^2 + R)t]/\sqrt{2\pi}$. This differs from the previous case only by inclusion of the factor $\exp(-Rt)$, which is independent of k. Calculation of the inverse Fourier transform proceeds as before, and we obtain

$$K(x,t) = \frac{\exp(-Rt)\exp(-x^2/4Dt)}{\sqrt{4\pi Dt}}. \quad \blacksquare$$

The following example and exercises provide practice in obtaining the concentration from a convolution integral.

4.2 Diffusion in an infinite medium

Example 4.2

The error function erf(x) was defined in Chapter 1 (Subsection 1.3.1) by the integral

$$\text{erf}(x) = \frac{2}{\sqrt{\pi}} \int_0^x dy \exp(-y^2).$$

Show that the concentration which results from an initial distribution at $t = 0$ which is c_0 for $x \geq a$, and 0 for $x < a$, is (for $t > 0$)

$$c(x,t) = \tfrac{1}{2} c_0 \left[1 + \text{erf}\left(\frac{x-a}{\sqrt{4Dt}}\right)\right].$$

Solution

Use equations (4.7) and (4.8):

$$c(x,t) = \frac{1}{\sqrt{4\pi Dt}} \int_{-\infty}^{\infty} dx_0 \exp[-(x-x_0)^2/4Dt] \, c(x_0, 0)$$

$$= \frac{c_0}{\sqrt{4\pi Dt}} \int_a^{\infty} dx_0 \exp[-(x-x_0)^2/4Dt],$$

where the properties of the initial distribution, $c(x_0, 0)$, were used in the second equality. Using the change of variable $u = (x - x_0)/\sqrt{4Dt}$, this integral is expressed in terms of an error function, plus a standard Gaussian integral:

$$c(x,t) = -\sqrt{4Dt}\, \frac{c_0}{\sqrt{4\pi Dt}} \int_{(x-a)/\sqrt{4Dt}}^{-\infty} du \exp(-u^2)$$

$$= \frac{c_0}{\sqrt{\pi}} \int_{-\infty}^{(x-a)/\sqrt{4Dt}} du \exp(-u^2)$$

$$= \frac{c_0}{\sqrt{\pi}} \int_{-\infty}^{0} du \, \exp(-u^2) + \frac{c_0}{\sqrt{\pi}} \int_0^{(x-a)/\sqrt{4Dt}} du \exp(-u^2)$$

$$= \frac{c_0}{2} + \frac{c_0}{2} \text{erf}\left(\frac{x-a}{\sqrt{4Dt}}\right).$$

Note that the first integral in the penultimate line is equal to one half of the standard Gaussian integral discussed in Block 0, Chapter 1, because its integrand is an even function. Figure 4.1 shows the concentration for various times for this initial condition, in the case where $D = 1$, $c_0 = 1$ and $a = 0$. We see that as time increases, diffusion causes particles to move further into the initially empty region $x < 0$. ∎

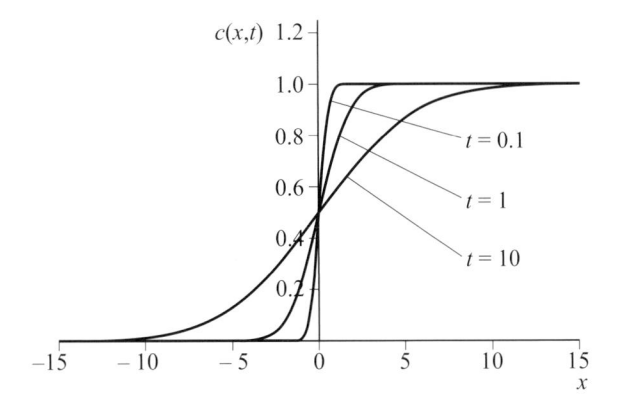

Figure 4.1 Concentration at successive times when the initial concentration is $c_0 = 1$ for $x \geq 0$ and zero for $x < 0$. The diffusion constant is $D = 1$.

Exercise 4.3

Determine $c(x,t)$ if the concentration at time $t=0$ is c_0 for $-L \leq x < L$, and zero elsewhere.

Harder exercise

A good way to tackle this problem is to use the result of Example 4.2, together with the fact that the diffusion equation is linear. Proceed as follows.

(a) Introduce a function $H(x)$ which is zero when $x < 0$ and unity when $x \geq 0$. Express the initial condition of Example 4.2 in terms of this function.

(b) Now express the initial condition of this problem as a sum of two terms of the form $b\, H(x-a)$, with different values of a and b.

(c) Use linearity to write the solution at time t.

This function is called the Heaviside function or step function; it is useful in many areas of applied mathematics.

Figure 4.2 shows the concentration for various times for the initial condition of Exercise 4.3. The diffusion causes particles to spread away from the region in which they are initially concentrated.

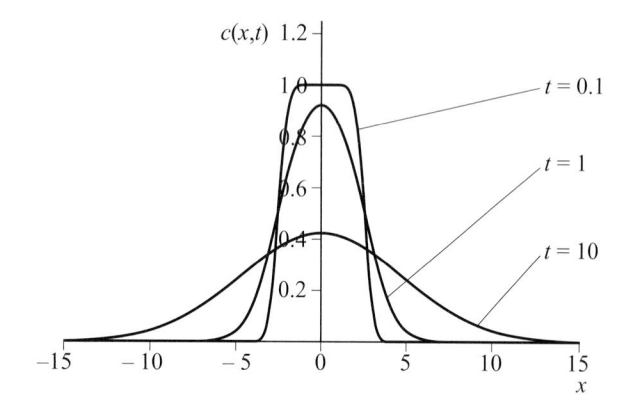

Figure 4.2 Concentration at successive times when the initial distribution is a 'top hat' function. In this example the initial concentration is unity on the interval $[-2.5, 2.5]$, and the diffusion constant is $D=1$.

Exercise 4.4

Determine the concentration at time t if the initial concentration at time $t=0$ is $c_0(x) = A_0 + A_1 \cos(kx)$. (Assume $A_0 > A_1 > 0$, so that the concentration is never negative.)

[Hint: You may assume the integral identity

$$\int_{-\infty}^{\infty} dx \exp(-\alpha x^2) \cos(\beta x) = \sqrt{\frac{\pi}{\alpha}} \exp(-\beta^2/4\alpha),$$

and recall that $\cos(A-B) = \cos A \cos B + \sin A \sin B$.]

Note that the amplitude of the oscillations in the concentration decrease exponentially with time, decreasing by a factor e when the time increases by $\tau = 1/(Dk^2)$.

In both of the plots above (Figures 4.1 and 4.2), we set $D=1$. If we want to understand the form of the solutions for other values of D, these plots are still informative. Note that the diffusion constant and time always occur in the solutions as the product Dt, so increasing D has the same effect as decreasing t by the same factor. For example, if the diffusion constant were $D=10$, the same curves would be obtained in Figures 4.1 and 4.2, but with the time labels changed to $t=0.01$, $t=0.1$ and $t=1$.

4.3 Solution in a finite medium in one dimension

Many problems involving the diffusion equation concern situations in which diffusion occurs in a finite region. A good example is the flow of heat along a metal rod, for which the temperature satisfies the diffusion equation. Let us consider a rod with length L and a uniform cross-section, with the thickness of the rod being small compared to L. The coordinate x measures distance along the rod, with $x = 0$ and $x = L$ being the ends of the rod (see Figure 4.3). The distribution of temperature θ in a homogeneous system has already been shown to satisfy the three-dimensional diffusion equation. If the rod is *thermally insulated* along its length so that no heat enters or leaves it (except at its ends), the temperature rapidly becomes very nearly uniform across the cross-section of the rod, so that the temperature distribution at time t is a function of just one coordinate, the distance x along the rod. It takes a much longer time for the entire rod to reach the same temperature; during this time the temperature satisfies a one-dimensional diffusion equation (because the derivatives of θ with respect to y and z are negligible).

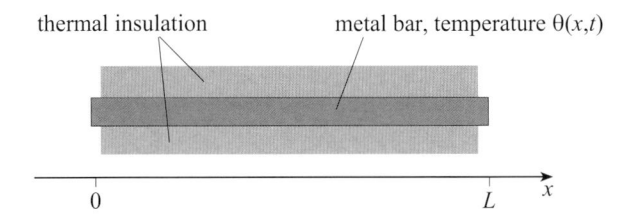

Figure 4.3 The temperature distribution $\theta(x, t)$ of a thermally insulated metal bar obeys the one-dimensional diffusion equation

> Heat loss can be reduced by surrounding the bar with a material with very low thermal conductivity, such as polystyrene foam. In practice, the one-dimensional diffusion equation is also a fair model for a heavy rod (such as a poker), made from a metal such as steel which is a good conductor of heat, suspended in air, as illustrated in Figure 4.4. This rod would be described as being *thermally isolated*, because the air has little capacity to absorb and carry away heat, rather than thermally insulated.

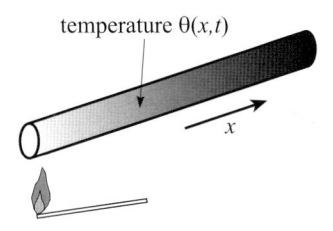

Figure 4.4 A metal rod is heated non-uniformly, and we follow how the temperature at different positions changes *after the heat source is removed*. The temperature in a long, thin, thermally isolated rod rapidly becomes uniform across its cross-section. The temperature distribution $\theta(x, t)$ then obeys the one-dimensional diffusion equation.

> This problem of determining how the temperature of a solid body varies with position and time has an important place in the history of mathematics. Fourier series were originally invented to solve just this problem, the flow of heat in a finite system. Fourier's work was rather more general than our approach,

because he also treated loss of heat from the rod to its surroundings, using the partial differential equation introduced in Example 4.1.

Now we consider the method for solving the diffusion equation in this finite system. The temperature at position x along a thermally insulated bar is initially $\theta_0(x)$ at time $t = 0$. The ends of the bar are assumed to be at $x = 0$ and at $x = L$. The aim is to calculate the time dependence of the temperature for each position along the rod, described by a function of two variables, $\theta(x, t)$.

We start by describing all of the information we have about the problem, as illustrated in Figure 4.4, in mathematical terms. We shall assume that the initial temperature of the rod at time $t = 0$ is known, and that it is given by a function $\theta_0(x)$. We know that the temperature distribution satisfies the one-dimensional diffusion equation. Also, because the bar is thermally isolated, there is no heat flow at the ends of the bar, at $x = 0$ and $x = L$. Because the heat flow is proportional to the temperature gradient, $\partial \theta / \partial x$ is equal to zero at those points. The temperature distribution $\theta(x, t)$ therefore satisfies

$$\left.\begin{aligned} \frac{\partial \theta}{\partial t} &= D \frac{\partial^2 \theta}{\partial x^2} \quad \text{(diffusion equation, valid for } t > 0\text{)}, \\ \frac{\partial \theta}{\partial x}(0, t) &= \frac{\partial \theta}{\partial x}(L, t) = 0 \quad \text{(boundary conditions, valid for } t > 0\text{)}, \\ \theta(x, 0) &= \theta_0(x) \quad \text{(initial condition at } t = 0\text{)}. \end{aligned}\right\} \quad (4.10)$$

The statement of the problem in equations (4.10) thus involves three elements: these are the *partial differential equation*, the *boundary conditions*, and the *initial conditions*.

Partial differential equations are difficult to solve, and it is a good idea to attempt to simplify the problem by reducing it to ordinary differential equations. We use the method of separation of variables. This was introduced in Block I, Chapter 2 as a method for solving the wave equation, but you will see that the same procedure works for the diffusion equation.

4.3.1 Separation of variables

We write $\theta(x, t)$ as a product

$$\theta(x, t) = f(x)\, g(t), \tag{4.11}$$

where f and g are two functions which are to be determined. Equation (4.11) is substituted into the diffusion equation in order to discover what must be required of the unknown functions f and g. The resulting equation can be arranged in the form

$$\frac{g'(t)}{g(t)} = D \frac{f''(x)}{f(x)}. \tag{4.12}$$

Primes denote derivatives: $f'(x) = df/dx$, etc.

Note that the left-hand side of equation (4.12) is a function of t only, but the right-hand side is a function of x only. These statements imply that both sides of the equation are constant, independent of both x and t. This constant is called the *separation constant*. We shall initially assume that the separation constant must be negative or zero, and it is convenient to write it as $-Dk^2$, where k is a real number to be determined. The functions f and g therefore satisfy the independent ordinary differential equations

$$f'' + k^2 f = 0 \tag{4.13}$$

and

$$g' + Dk^2 g = 0. \tag{4.14}$$

For $k \neq 0$, these equations have solutions $f(x) = A\cos(kx) + B\sin(kx)$ (where A and B are constants) and $g(t) = \exp(-Dk^2 t)$. In the special case where $k = 0$, we have $f(x) = A + Bx$, and $g = 1$. Note that because $f(x)$ and $g(t)$ occur as a product in solution (4.11) for $\theta(x, t)$, the arbitrary constant of integration coming from equation (4.14) for $g(t)$ is absorbed into the constants A and B appearing in the solution for $f(x)$.

4.3.2 Identifying eigenfunctions

Having identified the functions $f(x)$ and $g(t)$, we now consider the boundary conditions. The solution $\theta(x, t) = f(x)g(t)$ should satisfy $\partial \theta / \partial x = 0$ at $x = 0$ and at $x = L$. This implies that the derivative $f'(x)$ must equal zero at these points: that is, the function $f(x)$ should satisfy the boundary conditions $f'(0) = f'(L) = 0$. The functions $f(x)$ satisfying both the differential equation (4.13) and the boundary conditions are called eigenfunctions, and the corresponding values of k^2 are the eigenvalues of equation (4.13).

The condition $f'(0) = 0$ can be satisfied only if $B = 0$, so $f(x) = A\cos(kx)$. The condition $f'(L) = 0$ is then satisfied by choosing k such that $kL = n\pi$, with $n = 0, 1, 2, 3, \ldots$. The eigenfunctions are therefore $f_n(x) = A\cos(n\pi x/L)$, and the eigenvalues are $k_n^2 = (n\pi/L)^2$. Note that this solution is valid when $n = 0$, so that $k = 0$ and $f(x) = A$, corresponding to the special case ($k = 0$) mentioned above.

The method of separation of variables thus leads to a set of solutions of the form
$$\theta(x,t) = A_n \cos(k_n x) \exp(-Dk_n^2 t), \quad k_n = \frac{n\pi}{L}, \quad n = 0, 1, 2, \ldots, \quad (4.15)$$
where each A_n is an undetermined constant, and we have used the expression for $g(t)$ quoted below equation (4.14).

We should briefly consider whether other acceptable solutions could be obtained by taking a positive separation constant, $+Dk^2$, say. This choice would lead to a solution of equation (4.14) of the form $\exp(Dk^2 t)$, which is exponentially increasing, contrary to expectations from everyday experience (and the laws of physics), which suggest that temperature differences between different parts of the rod should decrease. A more compelling reason for rejecting the positive separation constant comes from considering the form of the spatial dependence, which becomes $f(x) = a\cosh(kx) + b\sinh(kx)$. The boundary condition $f'(0) = f'(L) = 0$ cannot be satisfied for any value of k other than $k = 0$ (apart from in the trivial case where $a = b = 0$). The case $k = 0$ corresponds to $f(x) = \text{constant}$, which is the $n = 0$ case of the set of solutions given in equations (4.15).

4.3.3 General solution

Because the diffusion equation is a linear equation, a general solution is obtained as a linear combination of the solutions (4.15). A general solution of the diffusion equation satisfying the boundary conditions for $t \geq 0$ is
$$\theta(x,t) = \sum_{n=0}^{\infty} A_n \cos\left(\frac{\pi n x}{L}\right) \exp\left(-\frac{\pi^2 n^2 D t}{L^2}\right). \quad (4.16)$$

Note that the exponential factors $\exp(-Dk_n^2 t)$ result in all of the temperature differences decreasing over time, with the components with more rapid spatial variation (those with larger values of k_n) decreasing more rapidly. The same exponential factor was seen earlier, in Exercise 4.4.

Exercise 4.5

Check that equation (4.16) does satisfy both the diffusion equation and the boundary conditions at both $x = 0$ and $x = L$.

4.3.4 Orthogonality relation

The coefficients A_n in equation (4.16) remain to be determined. This is done by choosing them so that the initial condition at $t = 0$ is satisfied. Setting $t = 0$ in equation (4.16) gives an expression for $\theta_0(x) = \theta(x, 0)$ in the form

$$\theta_0(x) = \sum_{n=0}^{\infty} A_n \cos\left(\frac{\pi n x}{L}\right). \tag{4.17}$$

This can be recognised as a Fourier series, of the form introduced in Block I, Chapter 3, equation (3.1); here the period is $2L$, and all of the b_n coefficients are zero. Formulae for determining the Fourier coefficients using orthogonality relations were discussed in Block I, Chapters 2 and 3. For this Fourier series, we have the orthogonality relation

$$\frac{2 - \delta_{n0}}{L} \int_0^L dx \cos\left(\frac{n\pi x}{L}\right) \cos\left(\frac{m\pi x}{L}\right) = \delta_{nm}. \tag{4.18}$$

(The multiplier is $2/L$ when $n \neq 0$ and $1/L$ when $n = 0$; the Kronecker delta symbol is used here so that the case $n = 0$ can be covered without writing a separate equation.) This can be checked by calculating the integral directly, but Section 4.4 will discuss a more direct and more fundamental approach to demonstrating orthogonality relations.

4.3.5 Calculation of Fourier coefficients

We can now calculate the Fourier coefficients in equation (4.17). We multiply both sides by $f_m(x) = \cos(m\pi x/L)$, integrate, and use the orthogonality relation (4.18):

$$\int_0^L dx\, \theta_0(x) f_m(x) = \sum_{n=0}^{\infty} A_n \int_0^L dx\, \cos(n\pi x/L) \cos(m\pi x/L)$$

$$= \sum_{n=0}^{\infty} A_n \frac{L}{2 - \delta_{n0}} \delta_{nm} = \frac{L A_m}{2 - \delta_{m0}}. \tag{4.19}$$

So (replacing the index m with n) the Fourier coefficients are

$$A_n = \frac{2 - \delta_{n0}}{L} \int_0^L dx \cos\left(\frac{\pi n x}{L}\right) \theta_0(x). \tag{4.20}$$

Having determined the coefficients A_n, we have a solution (equation (4.16)) which satisfies all of the requirements listed in equations (4.10). The solution is in the form of an infinite series. We shall not consider the convergence of this series in detail, but we note that for any positive value of t, the factors $\exp(-\pi^2 n^2 Dt/L^2)$ in equation (4.16) make the terms decrease very rapidly as $n \to \infty$. Therefore, for positive times, the series in equation (4.16) always converges nicely.

4.3 Solution in a finite medium in one dimension

Exercise 4.6

Obtain the coefficients A_n in the case where $\theta_0(x) = x$ on the interval $0 < x < L$.
[Hint: A similar Fourier series was obtained in Block I, Chapter 3, Exercise 3.6.]

Exercise 4.7

A steel poker of length L has one end in a fire and the other end in a bowl of water. It reaches a steady state, in which $\partial\theta/\partial t = 0$. Assuming that the temperature obeys the diffusion equation, show that the general form for $\theta(x)$ in this case is $\theta(x) = \bar{\theta} + \alpha(x - \frac{1}{2}L)$, where $\bar{\theta}$ and α are constants. Note that $\bar{\theta}$ is the mean value of the temperature of the poker.

[Hint: This is nothing to do with Fourier series: just write down and solve the diffusion equation for the steady state, with $\partial\theta/\partial t = 0$.]

Exercise 4.8

The poker in the previous exercise is removed from the 'source' and 'sink' of heat (i.e. the fire and the water, respectively), and is immediately placed in a thermally insulating environment. Show that the time-dependence of the temperature from the moment the poker becomes thermally isolated is given by

$$\theta(x,t) = \bar{\theta} - \sum_{n=1}^{\infty} \frac{2\alpha L}{\pi^2} \frac{[1-(-1)^n]}{n^2} \exp\left(-\frac{n^2\pi^2 Dt}{L^2}\right) \cos\left(\frac{n\pi x}{L}\right).$$

Note that only terms with n odd contribute to the sum. Use the substitution $n = 2k + 1$ to re-write the summation in a form with no terms which are identically zero.

The temperature of the poker in Exercise 4.8 is plotted in Figure 4.5 for three different times.

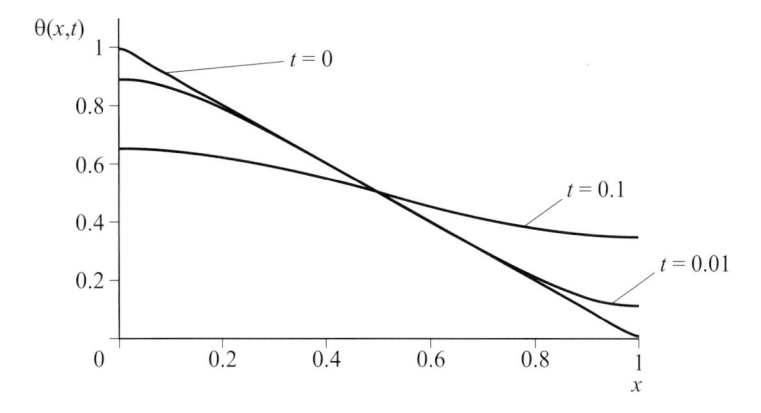

Figure 4.5 The temperature in a thin, thermally isolated metal rod, which initially has a uniform temperature gradient, where the thermal diffusion constant is $D = 1$, length is $L = 1$, temperature gradient is $\alpha = -1$, and average temperature is $\bar{\theta} = \frac{1}{2}$

The example of a metal rod with an initially uniform temperature gradient illustrates a subtle point about the representation of functions by Fourier series. In this example, the temperature gradient at the ends of the rod is initially α. However, the temperature $\theta_0(x)$ was represented by a Fourier series in the form of a sum of cosines, each term of which has a derivative which vanishes at $x = 0$ and at $x = L$. It therefore appears that the infinite sum of the Fourier series can have a property which is not present in any of its component terms, or any finite sum of the first N terms, no matter how large we make N. Understanding such subtleties of the convergence of Fourier series was a stimulus to refine pure mathematical analysis in the nineteenth century.

As soon as the rod is made thermally isolated, there is no flux of heat at its ends. At any instant after the rod has been disconnected from the heat source and heat sink, the temperature gradient at the ends must be zero. This effect is clearly visible in Figure 4.5, where the solution after the very short time $t = 0.01$ is very close to the $t = 0$ initial condition, except at the ends of the rod.

4.4 Orthogonality of eigenfunctions

You may have encountered the term 'eigenvalue' when studying matrices, and the term 'orthogonal' may be familiar from studying geometry or vectors. This section explains the connection between the older uses of these terms and their use in discussions of solutions of differential equations.

In Section 4.3 and several places in Block I we have seen that eigenfunctions of differential equations have an orthogonality property, which proves very convenient when determining the coefficients of the general solution. Until now, these orthogonality relations have been discussed on a case-by-case basis, and you may have wondered whether there is any general principle which implies that the eigenfunctions are always orthogonal. This section shows how the orthogonality relation discussed in Section 4.3 can be obtained without calculating an integral. The method can be adapted to many other problems involving solving ordinary and partial differential equations. In fact, it is often possible to show that the sets of eigenfunctions are orthogonal, even in cases where they cannot be calculated exactly.

4.4.1 Eigenfunctions and eigenvalues

Although the term eigenvalue may already be familiar from studying matrices, in greater generality, eigenvalues are properties of linear operators, which include matrices.

An operator, denoted by A, takes a function f and maps it to another function, $g = Af$. A linear operator A has the property that if f_1, f_2 are two functions, and α_1, α_2 are two constants, then

$$A(\alpha_1 f_1 + \alpha_2 f_2) = \alpha_1 A f_1 + \alpha_2 A f_2. \tag{4.21}$$

An example of a linear operator acting on a function $f(x)$ is differentiation: Af is the function df/dx. Another example is multiplication by another function $a(x)$, so that Af is the function $a(x)f(x)$.

An eigenfunction of A is a function f which satisfies

$$Af = \lambda f \tag{4.22}$$

for some constant λ, which is called an eigenvalue. We say that f is an eigenfunction of A associated with the eigenvalue λ. Often, additional conditions (such as boundary conditions) are imposed upon the function f.

An $N \times N$ square matrix $\mathbf{A}$ can be regarded as a linear operator, in the sense described above. In this case the functions f correspond to vectors. A vector $\boldsymbol{x}$ with N elements x_i, $i = 1, \ldots, N$, can be regarded as a mapping or function from the ordered set $\{1,\ldots,N\}$ to the ordered set $\{x_1,\ldots,x_N\}$. The eigenvalues λ and eigenvectors $\boldsymbol{x}$ of the matrix are defined by

$$\mathbf{A}\boldsymbol{x} = \lambda \boldsymbol{x}. \tag{4.23}$$

This is a special case of the previous equation, so eigenvectors of matrices are simply a special case of eigenfunctions of linear operators.

4.4.2 Orthogonality of functions

Now let us consider why two functions are said to be orthogonal if the integral of their product is equal to zero. Two lines are said to be orthogonal if they are at right angles. If these two lines are represented by the vectors $\boldsymbol{a}$ and $\boldsymbol{b}$ (i.e. vectors aligned along the directions of the lines), the condition for the lines to be orthogonal is that the scalar product of the two vectors is zero:

$$\boldsymbol{a} \cdot \boldsymbol{b} = \sum_{i=1}^{3} a_i b_i = 0, \tag{4.24}$$

where the a_i and b_i are the components of the vectors $\boldsymbol{a}$, and $\boldsymbol{b}$ respectively.

Two functions f and g are said to be orthogonal (on the interval $[0, L]$) if

$$\int_0^L dx\, f(x)\, g(x) = 0. \tag{4.25}$$

When dealing with linear operators, it is often helpful to think of the functions they act upon as being vectors in an infinite-dimensional space. The analogy with equation (4.24) becomes clear if the integral is approximated by a sum. Introduce two vectors $\boldsymbol{f}_\epsilon$ and $\boldsymbol{g}_\epsilon$ of dimension $N = \text{Int}(L/\epsilon)$ (i.e. N is the integer part of L/ϵ) with elements

$$f_j = \sqrt{\epsilon}\, f(j\epsilon), \quad g_j = \sqrt{\epsilon}\, g(j\epsilon), \quad j = 1,\ldots,N, \tag{4.26}$$

so

$$\boldsymbol{f}_\epsilon \cdot \boldsymbol{g}_\epsilon = \epsilon \sum_{j=1}^{N} f(j\epsilon)\, g(j\epsilon). \tag{4.27}$$

Then, from the usual definition of an integral as a limit, we have

$$\int_0^L dx\, f(x)\, g(x) = \lim_{\epsilon \to 0} \boldsymbol{f}_\epsilon \cdot \boldsymbol{g}_\epsilon, \tag{4.28}$$

so the integral in equation (4.28) may be thought of as a scalar product of two infinite-dimensional vectors. It is then natural to define functions to be orthogonal if the integral of their product vanishes.

4.4.3 Orthogonality relation

In Section 4.3 we derived the eigenfunctions $f_n(x)$ which arise when finding the temperature distribution in a thermally isolated rod. If we normalise these eigenfunctions such that

$$\int_0^L dx\, [f_n(x)]^2 = 1, \tag{4.29}$$

the normalised eigenfunctions are

$$f_n(x) = \sqrt{\frac{2-\delta_{n0}}{L}} \cos\left(\frac{n\pi x}{L}\right), \quad n = 0, 1, 2, \ldots. \tag{4.30}$$

These eigenfunctions are orthonormal, and satisfy

$$\int_0^L dx\, f_n(x)\, f_m(x) = \delta_{nm}, \tag{4.31}$$

which is equivalent to equation (4.18). This can be checked by calculating the integral directly, but there is an alternative method, which allows us to demonstrate that eigenfunctions of certain differential equations are orthogonal, even when we cannot write down formulae for the solutions. (This approach will be developed further in Section 4.5.) The idea is to obtain the orthogonality relation directly from the differential equation and boundary conditions satisfied by $f(x)$, without having to evaluate the integral. Let $f_n(x)$ and $f_m(x)$ be solutions of the differential equation, such that $f_n'' + k_n^2 f_n = 0$, $f_n'(0) = f_n'(L) = 0$, and f_m satisfies the same relations with a different separation constant $k_m^2 \neq k_n^2$. Now consider an integral I_{nm} defined by

$$I_{nm} = \int_0^L dx\, [f_n(x) f_m''(x) - f_m(x) f_n''(x)]. \tag{4.32}$$

We note that

$$\frac{d}{dx}\left[f_n(x) f_m'(x) - f_m(x) f_n'(x)\right] = f_n(x) f_m''(x) - f_m(x) f_n''(x), \tag{4.33}$$

so we can integrate equation (4.32) to obtain

$$I_{nm} = \left[f_n(x) f_m'(x) - f_m(x) f_n'(x)\right]_0^L. \tag{4.34}$$

The boundary conditions imply that $I_{nm} = 0$, because $f_n'(x)$ and $f_m'(x)$ are zero at $x = 0$ and $x = L$. We can also simplfy the expression for I_{nm} in equation (4.32) by using the differential equation to write $f_n'' = -k_n^2 f_n$, and similarly for f_m, so equation (4.32) becomes

$$I_{nm} = (k_n^2 - k_m^2) \int_0^L dx\, f_n(x)\, f_m(x) = 0. \tag{4.35}$$

We conclude that if $k_n \neq k_m$, the solutions f_n and f_m are orthogonal for the interval on which the differential equation is considered. This approach will work in many other problems involving linear partial differential equations, enabling us to show that eigenfunctions are orthogonal, without doing a detailed calculation. The case where $n = m$ in equation (4.31) must be checked separately by a direct calculation of the integral.

4.5 Diffusion in a finite medium in two and three dimensions

Next we consider how to treat diffusion of heat or particles in a finite medium in two and three dimensions: the same methods are applicable in both cases, but we must interpret symbols representing vectors as two-dimensional ($r = (x, y)$) or three-dimensional ($r = (x, y, z)$) objects according to the context. We discuss the problem in terms of the flow of heat in a *thermally isolated* system, and solve the diffusion equation for the temperature $\theta(r, t)$. Exactly the same approach is applicable to describing the diffusion of particles using a concentration $c(r, t)$. We shall follow the method used in the one-dimensional case as closely as possible. You will see that this leads to an interesting generalisation of the notion of Fourier series.

4.5.1 The Helmholtz equation

Consider the problem of determining the temperature distribution inside a thermally isolated piece of material in three dimensions. The initial temperature distribution is assumed to be known. The boundary-value problem now takes the form

$$\left.\begin{aligned}
\text{differential equation:} \quad & \tfrac{\partial \theta}{\partial t} = D\nabla^2 \theta, \quad t > 0; \\
\text{boundary condition:} \quad & \nabla\theta \cdot \hat{\mathbf{n}} = 0, \quad t > 0 \text{ and } r \text{ on the boundary}; \\
\text{initial condition:} \quad & \theta(r, 0) = \theta_0(r).
\end{aligned}\right\} \quad (4.36)$$

These equations are a natural extension of the one-dimensional case, equations (4.10). The first line specifies that $\theta(r, t)$ obeys the diffusion equation for all $t > 0$ and all positions inside the volume V.

The second line of equations (4.36) specifies the boundary condition that the gradient of the solution in the direction of the normal to the boundary is zero. (Here $\hat{\mathbf{n}}$ is the unit vector normal to the boundary.) Unless otherwise stated, we assume that the boundary of the region is smooth, so the outward normal $\hat{\mathbf{n}}$ is well defined. This condition holds for all positions on the boundary, and for all positive times. This boundary condition is a consequence of the system being thermally isolated (that is, not transferring heat to or from its surroundings). If no heat flows to or from the surroundings, the flux of heat across every element of the surface is zero, and Fourier's law (Chapter 3, Section 3.6) implies that the gradient of the temperature in the direction normal to the surface is zero everywhere on the surface. Sometimes the notation $\partial \theta / \partial n$ is used for this 'normal gradient'.

The third line of equations (4.36) specifies the initial condition that $\theta = \theta_0(r)$ at time $t = 0$. (The initial distribution, $\theta_0(r)$, is arbitrary, and the boundary condition need not be satisfied by the initial condition in cases where heat has been flowing into a region which is suddenly made thermally isolated at time $t = 0$. A one-dimensional version of this situation was exemplified in Exercise 4.8.)

The solution follows the one-dimensional case in that it uses the method of separation of variables, but there are some differences. The first step is to separate variables, writing a trial solution in the form

$$\theta(r, t) = g(t)\, f(r). \quad (4.37)$$

Substitution into the diffusion equation, and repeating the argument for separation into two independent equations, shows that f satisfies an equation analogous to (4.13), of the form

$$\nabla^2 f + \lambda f = 0, \tag{4.38}$$

where λ is a real constant. This equation is known as the *Helmholtz equation*. The function $g(t)$ satisfies the equation $g' + D\lambda g = 0$, as before, except that we have changed the name of the separation constant from k^2 to λ. The Helmholtz equation also arises in solving the three-dimensional wave equation, after separating out the time-dependent part of the solution.

Exercise 4.9

Derive equation (4.38) and the corresponding equation for $g(t)$.

> Hermann von Helmholtz (1821–94) obtained his equation by applying separation of variables to the wave equation. Helmholtz made lasting contributions to many areas of science, including the physiology of the eye, ear, larynx and nervous system, thermodynamics, fluid dynamics, acoustics, and electromagnetism.

We ensure that $\theta(\boldsymbol{r}, t)$ satisfies the boundary condition by requiring that solutions of the Helmholtz equation satisfy the boundary condition. That is, we require that the gradient of the function f in the direction normal to the boundary is zero. Mathematically, this is expressed by stating that $\boldsymbol{\nabla} f \cdot \widehat{\mathbf{n}} = 0$ for every point on the boundary, where $\widehat{\mathbf{n}}$ is a unit vector normal to the boundary at that point. This is called a *Neumann boundary condition*. It is a fact that solutions of the Helmholtz equation satisfying this boundary condition can be found only for a discrete set of non-negative values of λ. (This is hard to prove, but will be illustrated by an example later.) These values will be denoted by an integer subscript, labelling the λ_n as an ascending sequence, which starts from $n = 0$. There are an infinite number of these values λ_n for which solutions of the Helmholtz equation satisfying the appropriate boundary condition can be found. These values, λ_n, are known as the *eigenvalues* of the Helmholtz equation, and the corresponding functions $f_n(\boldsymbol{r})$ are called the *eigenfunctions*. The set of eigenvalues is called the *spectrum*.

Exercise 4.10

In the case of the one-dimensional example treated in Section 4.3, which equation corresponds to the Helmholtz equation? What are the eigenfunctions and eigenvalues?

4.5.2 A solution of the Helmholtz equation in two dimensions

There is no general technique for finding exact solutions of the Helmholtz equation in two or three dimensions which will always be successful. Solutions of the Helmholtz equation have been obtained by separation of variables in a limited number of cases. The approach works for systems with boundaries in the form of simple geometrical shapes (rectangular boundaries and those with circular or spherical symmetry), which are often the cases of most interest in applications. For a system with an irregularly shaped boundary, you would have to use a computer program to find numerical approximations to solutions of the Helmholtz equation.

> In two dimensions, the limitations on finding exact solutions of the Helmholtz equation have been studied in detail. Exact solutions are available for the rectangle, the circle, the ellipse, and three triangles (equilateral, 45–90–45 and 60–30–90). Even for a shape as simple as a triangle with other choices of angles, it is known that no exact expression for the eigenvalues of the Helmholtz equation can exist.

We now consider the solution of the Helmholtz equation by separation of variables for a rectangular domain in two dimensions, with the boundary condition $\nabla f \cdot \hat{\mathbf{n}} = 0$, where the sides of the rectangle have lengths a and b. A very similar problem in Block I discussed the solution of the two-dimensional Helmholtz equation arising from separation of variables for the wave equation.

We need to determine functions $f(x,y)$ which satisfy the Helmholtz equation

$$\frac{\partial^2 f}{\partial x^2} + \frac{\partial^2 f}{\partial y^2} + \lambda f = 0 \tag{4.39}$$

in the region $0 < x < a$, $0 < y < b$, with the boundary condition $\hat{\mathbf{n}} \cdot \nabla f = 0$. These eigenfunctions exist only for certain choices of λ, which are the eigenvalues of the problem.

Consider how to describe the boundary condition in a specific form. The boundary consists of four segments and four corners (illustrated in Figure 4.6), and the boundary condition takes a different form on each boundary component. The four segments (excluding the corners) are as follows.

Segment 1

The line segment $x = 0$, $0 < y < b$. The normal may be taken to be $\hat{\mathbf{n}} = -\mathbf{i}$. (The sign is arbitrary here, but we quote the choice of sign which makes $\hat{\mathbf{n}}$ point outside the rectangle.) The gradient vector ∇f is expressed in terms of partial derivatives of f, so the boundary condition $\nabla f \cdot \hat{\mathbf{n}} = 0$ becomes

$$\nabla f \cdot \hat{\mathbf{n}} = -\left(\frac{\partial f}{\partial x}\mathbf{i} + \frac{\partial f}{\partial y}\mathbf{j}\right) \cdot \mathbf{i} = -\frac{\partial f}{\partial x} = 0. \tag{4.40}$$

This condition must be satisfied at all points along the line segment. The boundary condition on this segment is therefore

$$\frac{\partial f}{\partial x}(0, y) = 0, \quad 0 < y < b. \tag{4.41}$$

Segment 2

The line segment $y = 0$, $0 < x < a$. On this segment $\hat{\mathbf{n}} = -\mathbf{j}$, and the boundary condition is $\partial f/\partial y = 0$. We therefore require

$$\frac{\partial f}{\partial y}(x, 0) = 0, \quad 0 < x < a. \tag{4.42}$$

Segment 3

The line segment $x = a$, $0 < y < b$. Here $\hat{\mathbf{n}} = \mathbf{i}$ and the boundary condition is the same as for segment 1, except that $x = a$:

$$\frac{\partial f}{\partial x}(a, y) = 0, \quad 0 < y < b. \tag{4.43}$$

Segment 4

The line segment $y = b$, $0 < x < a$. Here $\hat{\mathbf{n}} = \mathbf{j}$ and the boundary condition is

$$\frac{\partial f}{\partial y}(x, b) = 0, \quad 0 < x < a. \tag{4.44}$$

Corners

Corners pose a problem because $\hat{\mathbf{n}}$ is ill-defined on these. However, if $\boldsymbol{\nabla} f$ vanishes on these corners, then the appropriate (zero heat flux) boundary condition is satisfied. Consider two points very close to a corner, situated on either side of it, so that the normals are in two perpendicular directions. At each of these points the normal gradient vanishes, in two perpendicular directions. We make both points approach the corner, so that there is a point at which $\partial f/\partial x = 0$, which is arbitrarily close to a point at which $\partial f/\partial y = 0$. If we assume that the function $f(x, y)$ is twice differentiable, both components of $\boldsymbol{\nabla} f$ must approach zero (as required) as we approach the corner. You can check that the solutions we obtain do satisfy this criterion.

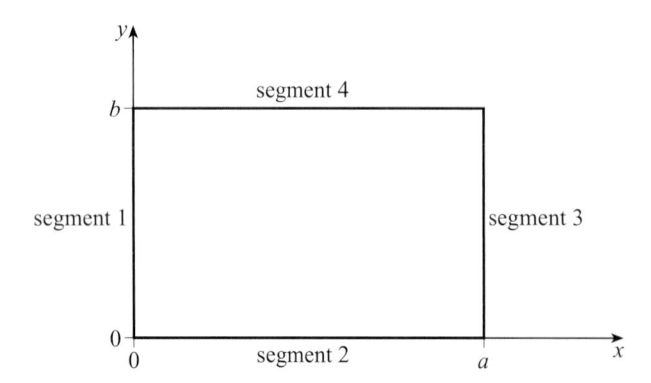

Figure 4.6 We consider the solution of the Helmholtz equation $(\nabla^2 + \lambda)f = 0$ on the interior of the rectangle $0 < x < a$, $0 < y < b$, satisfying the Neumann boundary condition $\boldsymbol{\nabla} f \cdot \hat{\mathbf{n}} = 0$. The boundary condition is considered separately for the different segments of the boundary.

We need solutions which satisfy the boundary condition everywhere on the boundary, implying that all four of these conditions must be satisfied. We guess that the method of separation of variables might work, and write

$$f(x, y) = X(x) Y(y). \tag{4.45}$$

Substituting this expression into the Helmholtz equation gives $X''Y + XY'' + \lambda XY = 0$. Dividing by XY gives

$$\frac{1}{X}\frac{d^2 X}{dx^2} = -\frac{1}{Y}\frac{d^2 Y}{dy^2} - \lambda. \tag{4.46}$$

The left- and right-hand sides of this relation are independent of y and x, respectively, and are therefore equal to a constant. Based on our experience with one-dimensional problems, we anticipate that negative or zero values of the new separation constant will produce sinusoidal solutions $X(x)$ which can satisfy the boundary conditions. If you wish, you can check that

4.5 Diffusion in a finite medium in two and three dimensions

the solutions found for a positive separation constant cannot satisfy the boundary conditions. We also expect that $Y(y)$ should have sinusoidal solutions. We write the separation constant for equation (4.46) as $-k_x^2$, and set $k_y = \sqrt{\lambda - k_x^2}$ (assuming that $\lambda - k_x^2 \geq 0$), so $\lambda = k_x^2 + k_y^2$. We then have to solve two equations:

$$X'' + k_x^2 X = 0, \qquad Y'' + k_y^2 Y = 0. \tag{4.47}$$

These equations have sinusoidal solutions, $X(x) = X_1 \cos(k_x x) + X_2 \sin(k_x x)$ and $Y(y) = Y_1 \cos(k_y y) + Y_2 \sin(k_y y)$, where X_1, X_2, Y_1, Y_2 are constants. In terms of the functions $X(x)$ and $Y(y)$, the boundary conditions are

$$\begin{aligned}
\text{segment 1:} &\quad X'(0)\ Y(y) = 0, \quad 0 \leq y \leq b; \\
\text{segment 2:} &\quad X(x)\ Y'(0) = 0, \quad 0 \leq x \leq a; \\
\text{segment 3:} &\quad X'(a)\ Y(y) = 0, \quad 0 \leq y \leq b; \\
\text{segment 4:} &\quad X(x)\ Y'(b) = 0, \quad 0 \leq x \leq a.
\end{aligned} \tag{4.48}$$

These conditions are satisfied by requiring that X and Y satisfy

$$X'(0) = X'(a) = 0, \qquad Y'(0) = Y'(b) = 0. \tag{4.49}$$

The functions X and Y therefore satisfy Neumann boundary conditions, as in the one-dimensional case (see page 122): the equation $X'' + k_x^2 X = 0$ is solved satisfying $X'(0) = X'(a) = 0$, and $Y'' + k_y^2 Y = 0$ is solved satisfying $Y'(0) = Y'(b) = 0$. Comparing with the one-dimensional case, L is replaced by a for $X(x)$ and by b for $Y(y)$. Thus, solutions for $X(x)$ and $Y(y)$ satisfying these boundary conditions are

$$X(x) = \cos\left(\frac{\pi n x}{a}\right), \qquad Y(y) = \cos\left(\frac{\pi m y}{b}\right), \tag{4.50}$$

where n, m are integers labelling the solutions. (We take n and m to be non-negative, to avoid giving two labels to the same solution.) The constants k_x, k_y are, respectively, $n\pi/a$ and $m\pi/b$. Because equations (4.47) are linear, these solutions can be multiplied by arbitrary constants.

Taking products of these functions, the eigenfunctions of the Helmholtz equation on the rectangle with Neumann boundary condition are therefore

$$f_{n,m}(x,y) = C_{n,m} \cos\left(\frac{n\pi x}{a}\right) \cos\left(\frac{m\pi y}{b}\right), \tag{4.51}$$

where each $C_{n,m}$ is an arbitrary constant. The corresponding eigenvalues of the Helmholtz equation for our rectangle are $\lambda = k_x^2 + k_y^2$. In terms of the two integers n, m labelling the eigenfunctions, these are

$$\lambda_{nm} = \left(\frac{n\pi}{a}\right)^2 + \left(\frac{m\pi}{b}\right)^2, \quad n = 0, 1, 2, \ldots, m = 0, 1, 2, \ldots. \tag{4.52}$$

In this example it was natural to use two integers n, m to label the eigenvalues λ. In cases where the Helmholz equation cannot be solved by separation of variables, the eigenvalues cannot be labelled in a natural way by pairs of integers. It is always possible to rank the eigenvalues by their size, and then label them by a single index, namely the order in this list. To account for this more general case, in some expressions below we shall use a single summation sign when we sum over eigenvalues.

Sometimes there are two or more distinct (that is, linearly independent) eigenfunctions $f(\mathbf{r})$ which have the same eigenvalue λ. In cases where there are two or more linearly independent eigenfunctions corresponding to a particular eigenvalue, that eigenvalue is said to be *degenerate*. If just two linearly independent eigenfunctions have the same eigenvalue, the eigenvalue is said to be *doubly degenerate*.

A set of N functions $f_i(x)$ is linearly independent if there are no (non-trivial) combinations of constants a_i such that $\sum_{i=1}^{N} a_i f_i(x) = 0$ for all x.

Exercise 4.11

Show that there are degeneracies in the spectrum of the Helmholtz equation with a Neumann boundary condition on the rectangle in the case where all the sides are of equal length (so that the region is square). Are there other cases where there are degeneracies?

Exercise 4.12

Show that the function $f(x) = \exp(ikx)$ is an eigenfunction of the differential operators d/dx and d^2/dx^2. If the function must satisfy the *periodic boundary condition* $f(x + a) = f(x)$ for all x, what are the eigenvalues and the spectrum of each of these operators? Are there any degeneracies?

4.5.3 Orthogonality and generalised Fourier series

Returning to the solution of the diffusion equation in two or three dimensions, a general solution satisfying the appropriate boundary condition may be constructed from the eigenfunctions and eigenvalues of the Helmholtz equation. Consider a linear combination of solutions of the form (4.37). The function $g(t)$ satisfies $g' + D\lambda g = 0$, which has a solution $g(t) = \exp(-D\lambda t)$. We therefore have solutions of the form

$$\theta(\boldsymbol{r}, t) = \sum_{n=0}^{\infty} a_n f_n(\boldsymbol{r}) \exp(-D\lambda_n t), \tag{4.53}$$

where the sum includes all eigenfunctions $f_n(\boldsymbol{r})$ and their associated eigenvalues λ_n, including cases where the eigenvalues are degenerate.

This solution satisfies the boundary conditions that were imposed upon the eigenfunctions $f_n(\boldsymbol{r})$ of the Helmholtz equation. The coefficients a_n are determined by the requirement that this solution reproduces $\theta_0(\boldsymbol{r})$ at time $t = 0$. It is not immediately clear how (or whether) this can be achieved. Answering this question requires a discussion of the properties of the eigenfunctions of the Helmholtz equation, which will lead to a powerful generalisation of the concept of Fourier series.

Note that here λ_n, etc., are labelled by a *single* index to account for more general situations.

In discussing properties of the $f_n(\boldsymbol{r})$, it will be assumed that the corresponding eigenvalues λ_n are distinct. This requirement is usually satisfied if the boundary has no symmetry properties, but it is not satisfied in many cases where there is a high degree of symmetry (such as the circular symmetry of a cylinder or if the boundary is a square or rectangle with rationally related side lengths). The eigenfunctions may be multiplied by any constant and remain solutions of the Helmholtz equation satisfying the Neumann boundary condition. We shall assume that the eigenfunctions are *real-valued* functions, and that the multiplying constant is chosen so that

$$\int_V dV \, [f_n(\boldsymbol{r})]^2 = 1, \tag{4.54}$$

where V is the region within which the function $f_n(\boldsymbol{r})$ satisfies the Helmholtz equation. Eigenfunctions satisfying equation (4.54) are said to be *normalised*. In the following discussion we shall assume that the eigenfunctions have been normalised.

We shall now show that the eigenfunctions also have an *orthogonality* property: they satisfy

$$\int_V dV \, f_n(\boldsymbol{r}) f_m(\boldsymbol{r}) = 0 \tag{4.55}$$

Note that the normalisation condition is different from that occurring in probability theory, because here we integrate the square of the function.

4.5 Diffusion in a finite medium in two and three dimensions

(provided $n \neq m$). This will be demonstrated using a generalisation of the argument given in Subsection 4.4.3. Consider the integral

$$I_{nm} = \int_V dV \, (f_n \nabla^2 f_m - f_m \nabla^2 f_n). \tag{4.56}$$

For any doubly differentiable scalar fields f and g, we have

$$\boldsymbol{\nabla} \cdot (f \boldsymbol{\nabla} g) = f \nabla^2 g + \boldsymbol{\nabla} f \cdot \boldsymbol{\nabla} g. \tag{4.57}$$

(This is an example of the result of Exercise 3.12, page 91, where we take $a = f$, $\boldsymbol{F} = \boldsymbol{\nabla} g$.) Using this result, we can write I_{nm} as an integral of the divergence of a vector field:

$$I_{nm} = \int_V dV \, \boldsymbol{\nabla} \cdot (f_n \boldsymbol{\nabla} f_m - f_m \boldsymbol{\nabla} f_n). \tag{4.58}$$

Using Gauss's theorem we can write I_{nm} as a surface integral over the surface S which is the boundary of the volume V:

$$I_{nm} = \int_S d\boldsymbol{S} \cdot (f_n \boldsymbol{\nabla} f_m - f_m \boldsymbol{\nabla} f_n) = 0. \tag{4.59}$$

Discussed in Block 0, Chapter 2, but see also Subsection 3.4.3 of this block.

The surface integral is equal to zero because of the boundary conditions which are satisfied by the eigenfunctions $f_n(\boldsymbol{r})$ and $f_m(\boldsymbol{r})$: the gradient of both of these functions in the direction $\hat{\boldsymbol{n}}$ normal to the surface is zero everywhere on the surface, and because $d\boldsymbol{S}$ is in the same direction as $\hat{\boldsymbol{n}}$, we have $\boldsymbol{\nabla} f_n \cdot d\boldsymbol{S} = 0$ everywhere on the surface.

Using the fact that both $f_n(\boldsymbol{r})$ and $f_m(\boldsymbol{r})$ satisfy the Helmholtz equation, the original expression for the integral I_{nm}, equation (4.56), can also be written

$$I_{nm} = (\lambda_n - \lambda_m) \int_V dV \, f_n(\boldsymbol{r}) \, f_m(\boldsymbol{r}). \tag{4.60}$$

Comparing equations (4.59) and (4.60) shows that the integral of the product $f_n(\boldsymbol{r}) f_m(\boldsymbol{r})$ over the region is zero when $\lambda_n \neq \lambda_m$. Functions satisfying this property are said to be *orthogonal*.

Assuming that all of the λ_n are distinct, and imposing the normalisation condition (4.54), the eigenfunctions $f_n(\boldsymbol{r})$ satisfy

$$\int_V dV \, f_n(\boldsymbol{r}) f_m(\boldsymbol{r}) = \delta_{nm} = \begin{cases} 1, & n = m, \\ 0, & n \neq m. \end{cases} \tag{4.61}$$

Sets of functions satisfying equation (4.61) are said to be *orthonormal*, that is, both orthogonal (for $n \neq m$) and normalised (for $n = m$). Note that this equation can be regarded as a generalisation of a result which was crucial in the development of Fourier series; see Block I, Chapter 3, equation (3.10).

> In Block I, Chapter 3, the functions $f_n(\boldsymbol{r})$ are *complex-valued* (as opposed to real-valued, one of the assumptions leading to equation (4.61)). In Block I, Chapter 3, the normalised functions are $f_n(x) = \exp(2\pi i n x / L)/\sqrt{L}$, and in order to deal with the complex values taken by $f_n(x)$, in equation (4.61) $f_m(x)$ is replaced by its complex conjugate, $f_m^*(x)$. Nothing more will be said about how the above discussion needs to be modified to allow for complex-valued functions $f_n(\boldsymbol{r})$, as in *this* section on higher-dimensional generalisations of Fourier series, *only* real-valued functions $f_n(\boldsymbol{r})$ will be encountered.

Now we return to the problem of determining a solution of the heat equation, which has been obtained in the form (4.53) with the Fourier coefficients a_n as yet undetermined. We shall assume that any function of interest (without specifying this set of functions precisely) can be expressed as a linear combination of the eigenfunctions $f_n(\boldsymbol{r})$. This property of the eigenfunctions is termed *completeness*. With this assumption, the initial condition is to be written as a linear combination of the $f_n(\boldsymbol{r})$:

$$\theta_0(\boldsymbol{r}) = \sum_{n=0}^{\infty} a_n f_n(\boldsymbol{r}). \qquad (4.62)$$

(This can be viewed as a generalisation of the Fourier series, as introduced in Block I, Chapter 3, equation (3.6), with the functions $f_n(\boldsymbol{r})$ replacing the exponential functions $\exp(2\pi i n x/L)/\sqrt{L}$.) To determine the Fourier coefficients a_n, we follow a procedure similar to that for the usual Fourier series: after relabelling the index n in equation (4.62) by m, we multiply equation (4.62) by $f_n(\boldsymbol{r})$ and integrate over the volume V:

$$\int_V dV \theta_0(\boldsymbol{r}) f_n(\boldsymbol{r}) = \sum_{m=0}^{\infty} a_m \int_V dV f_m(\boldsymbol{r}) f_n(\boldsymbol{r}). \qquad (4.63)$$

Equation (4.61) shows that the only integral in the sum which does not vanish is that for which $m = n$, and this integral is unity. It then follows that

$$a_n = \int_V dV \theta_0(\boldsymbol{r}) f_n(\boldsymbol{r}). \qquad (4.64)$$

This establishes the values of the coefficients a_n in equation (4.62). Substitution of these coefficients into equation (4.53) gives the solution of the diffusion equation, satisfying all of the conditions of equations (4.36).

The method described above followed the one-dimensional case quite closely, by introducing a broad generalisation of the idea of Fourier series, in which functions are expressed as linear combinations of eigenfunctions satisfying a differential equation and certain boundary conditions. Solutions based on generalised Fourier series are a powerful tool in further developments of the theory: they can be used to obtain interesting results even in cases where the eigenfunctions and eigenvalues cannot be calculated exactly.

The eigenfunctions can be shown to form an orthonormal set in situations where other boundary conditions are applicable. For example, if an object is placed in a stirred liquid with very high thermal conductivity, the temperature at the surface of the object will be constant (which may be taken to be zero). The appropriate boundary condition for the eigenfunctions is then $f_n(\boldsymbol{r}) = 0$ for all positions $\boldsymbol{r}$ on the boundary. This is called the *Dirichlet boundary condition*. The eigenfunctions are also orthogonal in this case.

Exercise 4.13

Show that the normalised eigenfunctions of the Helmholtz equation still satisfy equation (4.61) when the Dirichlet boundary condition ($f_n(\boldsymbol{r}) = 0$ for points $\boldsymbol{r}$ on the boundary) applies. Again, assume that there are no degenerate eigenvalues.

Exercise 4.14

Determine the normalised eigenfunctions of the rectangle with side lengths a and b using the Dirichlet boundary condition. Show that the eigenvalues are still given by equation (4.52), except that cases where $n = 0$ or $m = 0$ are excluded.

4.6 Semi-infinite domains: temperature waves

This section will consider a solution of the diffusion equation in a semi-infinite domain. (We describe the method of solution in terms of a heat flow problem.) Here we do not assume that the initial temperature distribution is known, but we do specify either the temperature or the heat flux at the boundary. The physical context is that of a surface which is alternately heated and cooled, with repeat period T. An example is the surface of the Earth, which is alternately heated and cooled during the day (in this case $T = 24$ hours). The temperature below this surface is expected to vary periodically as a function of time, also with period T. It is natural to expect that the amplitude of the oscillations in the temperature will become negligible at great depths below the surface, x, so we impose a boundary condition that the temperature approaches a constant value as the depth x approaches infinity. It is of interest to determine how the temperature varies as a function of depth, and how far the temperature variation lags behind the changes at the surface. If the surface is heated or cooled uniformly, we expect that the temperature will depend only on the depth x (and not on any other spatial coordinate) and time t, so (although this is a problem posed in three-dimensional space) we are concerned with finding a solution of the one-dimensional diffusion equation.

First consider the formulation of the problem in mathematical terms. The surface which is heated is the plane $x = 0$, and the soil is the half-space $x \geq 0$. We must determine a solution $\theta(x,t)$ satisfying the diffusion equation, and appropriate boundary conditions. We shall restrict ourselves to looking for solutions which vary periodically in time, with period T. (There can be non-periodic disturbances which decrease in amplitude as a function of time, often called the *transients*, which we shall not consider.) We apply the boundary condition that $\theta(x,t)$ approaches a constant ($\bar{\theta}$, say) as $x \to \infty$. It will be assumed that the heat flux density $J(t)$ (that is, the heat energy per unit time passing through a unit area perpendicular to the surface) at the surface varies sinusoidally:

$$J(t) = J_0 \sin(\omega t), \tag{4.65}$$

where $\omega = 2\pi/T$ is the angular frequency corresponding to the period T. (When J is negative, the surface is losing heat.) A general periodic variation of $J(t)$ is discussed later. The heat flux density is related to the temperature gradient at the surface by Fourier's law (discussed in Section 3.6),

$$J(t) = -\kappa \frac{\partial \theta}{\partial x}(0, t), \tag{4.66}$$

where κ is the thermal conductivity.

We expect that the temperature $\theta(x,t)$ varies with the same period as the changes at the surface. It is convenient to write the sinusoidal function as a sum of two complex exponentials: $\sin(\omega t) = [\exp(i\omega t) - \exp(-i\omega t)]/2i$. We then seek solutions of the one-dimensional diffusion equation $\partial \theta/\partial t = D\, \partial^2 \theta/\partial x^2$ in the form

$$\theta(x,t) = \exp(\pm i \omega t) f(x), \tag{4.67}$$

where $\pm$ indicates that we consider both plus and minus signs, corresponding to the two exponentials $\exp(\pm i\omega t)$ which make up the $\sin(\omega t)$ function.

Substitution of the trial solution (4.67) into the diffusion equation gives an ordinary differential equation for $f(x)$:

$$\pm i\omega f_{\pm} = D \frac{d^2 f_{\pm}}{dx^2}, \tag{4.68}$$

where $\pm$ indicates that we take a positive or negative sign to correspond with the the sign in $\exp(\pm i\omega t)$.

> Of course, our final solution to the problem must be a real-valued function, because the temperature measured by a thermometer is not a complex number. This will be achieved by combining two complex solutions to make a real solution.
>
> The motivation for using complex functions is that they make it easier to solve the problem, both by making it easier to spot a solution, and by simplifying the algebra. If you need to be convinced of this, try to re-work this problem without using complex functions. Even checking that our solution (equation (4.76) below) satisfies the diffusion equation is hard work without using complex exponentials.

Equation (4.68) has a solution of the form $f_+(x) = \exp(\alpha_+ x)$, where the subscript $+$ refers to the fact that the multiplier of ω in the left-hand side of equation (4.68) is $+i$, and where α_+ satisfies

$$i\omega = D\alpha_+^2 . \tag{4.69}$$

Rearranging this equation to determine α_+, we find

$$\alpha_+ = \pm\sqrt{\frac{i\omega}{D}} = \pm\sqrt{\frac{\omega}{2D}}(1+i) = \pm K(1+i), \tag{4.70}$$

where we use the fact that $(1+i)^2 = 2i$, so $\sqrt{i} = (1+i)/\sqrt{2}$, and where $K = \sqrt{\omega/2D}$. Note that there are two possible signs for the square root. The solution with the positive sign gives a function $f_+(x) = \exp(Kx)\exp(iKx)$, which grows as x increases. This solution must be rejected, because we assume that the influence of heating the surface becomes negligible at great depth. Only the solution with negative sign should be retained, which gives $f_+(x) = \exp(-Kx)\exp(-iKx)$ as an *acceptable* solution, and

$$\alpha_+ = -\sqrt{\frac{\omega}{2D}}(1+i) \tag{4.71}$$

as the acceptable value for α_+.

Changing the sign in front of ω in equation (4.68) results in a solution $f_-(x) = \exp(\alpha_- x)$, with α_- satisfying

$$-i\omega = D\alpha_-^2. \tag{4.72}$$

Again, the expression for α_- contains a square root, and one choice of sign leads to an unphysical solution, with $f_-(x)$ increasing as x increases. The second (acceptable) solution has

$$\alpha_- = -\sqrt{\frac{\omega}{2D}}(1-i), \tag{4.73}$$

where $(1-i)^2 = -2i$ has been used. The solution of the diffusion equation is a linear combination of the two acceptable solutions, after multiplying by

4.6 Semi-infinite domains: temperature waves

$\exp(i\omega t)$ or $\exp(-i\omega t)$ as appropriate:

$$\begin{aligned}\theta(x,t) &= A_+ \exp(i\omega t)f_+(x) + A_- \exp(-i\omega t)f_-(x) \\ &= A_+ \exp(i\omega t + \alpha_+ x) + A_- \exp(-i\omega t + \alpha_- x) \\ &= A_+ \exp\left[i\omega t - \sqrt{\frac{\omega}{2D}}(1+i)x\right] \\ &\quad + A_- \exp\left[-i\omega t - \sqrt{\frac{\omega}{2D}}(1-i)x\right]. \end{aligned} \quad (4.74)$$

The physical value of the temperature must be a real number. The solution (4.74) is real for all values of x and t if $A_+ = (A_-)^*$ (since then $\theta(x,t) = [\theta(x,t)]^*$, implying that $\theta(x,t)$ is real). Writing $A_\pm = \frac{1}{2}A\exp(\pm i\phi)$ gives a real solution of the form

$$\theta(x,t) = A\cos\left(\phi + \omega t - \sqrt{\frac{\omega}{2D}}\,x\right)\exp\left(-\sqrt{\frac{\omega}{2D}}x\right), \quad (4.75)$$

using $\cos u = (\exp(iu) + \exp(-iu))/2$. The cosine factor in this solution may be interpreted as a wave travelling in the positive x direction with speed $\sqrt{2D\omega}$. (Compare this with the travelling wave solutions considered in Block I.) The exponential factor in equation (4.75) indicates that the amplitude of this wave decreases with a 'scale length' $K^{-1} = \sqrt{2D/\omega}$.

> That the travelling wave has a speed proportional to $\sqrt{D\omega}$ can be interpreted through the following (approximate) argument. It is known that the (one-dimensional) diffusion equation has solutions corresponding to having the mean-squared distance, $\langle x^2 \rangle$, travelled by a particle over a time period T taking the value $\langle x^2 \rangle = 2DT$. (See, for example, Section 3.7 of the previous chapter, particularly equation (3.61) on page 98.) Thus, one can argue that the 'mean speed' of this diffusing particle is $\sqrt{\langle x^2 \rangle}/T = \sqrt{2D/T}$, which is proportional to $\sqrt{D\omega}$ because $T = 2\pi/\omega$. Since heat is carried by diffusing particles, one might, therefore, expect the temperature wave to travel at a speed proportional to $\sqrt{D\omega}$. Furthermore, note that the 'length scale', $K^{-1} = \sqrt{2D/\omega}$, is proportional to $\sqrt{2DT}$, the root-mean-squared distance, $\sqrt{\langle x^2 \rangle}$, travelled by a diffusing particle over the repeat period T. Thus, one might anticipate that the amplitude of the temperature wave diminishes on length scales larger than $\sqrt{2DT}$ since the diffusing particles (carrying heat from the surface) are less likely to get there.

The constants A and ϕ are determined by requiring that the boundary condition described by equations (4.65) and (4.66) is satisfied. The required solution is

$$\theta(x,t) = \frac{J_0}{\kappa}\sqrt{\frac{D}{\omega}}\cos\left(\omega t - \sqrt{\frac{\omega}{2D}}x + \frac{5\pi}{4}\right)\exp\left(-\sqrt{\frac{\omega}{2D}}x\right). \quad (4.76)$$

Exercise 4.15

Confirm this result, by showing that the choices $\phi = 5\pi/4$ and $A = (J_0/\kappa)\sqrt{D/\omega}$ make the solution (4.75) consistent with the boundary condition described by equations (4.65) and (4.66).

The solution (4.76) is illustrated in Figure 4.7. The temperature is plotted as a function of the depth below the surface, for four different times throughout the period of the heat flux density, $J(t)$. For this plot, we have chosen $D = \omega = J_0/\kappa = 1$. When these constants are assigned different values, the form of the curves remains the same (but of course the horizontal and vertical scales of the plots must change). Note that the temperature at points below the surface lags behind that at the surface, and that the amplitude of the temperature wave decreases rapidly with depth.

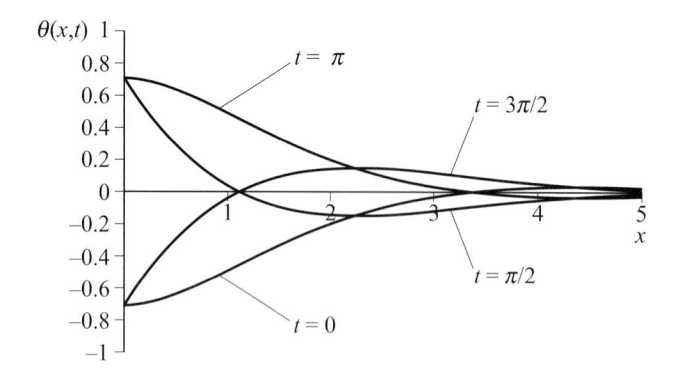

Figure 4.7 The temperature as a function of depth, for four different times, for the case $D = \omega = J_0/\kappa = 1$. The rate $J(t)$ at which heat is supplied is a maximum at $t = \pi/2$, and a minimum at $t = 3\pi/2$.

Exercise 4.16

What is the temperature at depth x if the temperature at the surface is $\theta_0 \sin(\omega t)$? [Hint: The solution is unchanged up to equation (4.75).]

Example 4.3

Generalise the calculation of the previous exercise to the case where the temperature $\theta_0(t)$ at the surface is an arbitrary periodic function of time, with period $T = 2\pi/\omega$. (Express $\theta_0(t)$ as a Fourier series, and use the fact that the diffusion equation is linear.)

What happens to the relative magnitudes of the terms in your expression as you go further from the surface?

Solution

Because the diffusion equation is linear, an arbitrary linear combination of its solutions also satisfies the diffusion equation. Any surface temperature which is a periodic function of time, with period T, may be written as a Fourier series:

$$\theta_0(t) = \frac{a_0}{2} + \sum_{n=1}^{\infty} \left[a_n \cos(n\omega t) + b_n \sin(n\omega t) \right],$$

where $\omega = 2\pi/T$ is the angular frequency corresponding to T. Adapting the results of the previous exercise by replacing ω with $n\omega$, and by allowing for a Fourier series containing both cosine and sine terms, we see that each component of this Fourier series corresponds to a solution of the diffusion equation of the form (4.75). Combining these solutions with appropriate values of A and ϕ, so that $\theta(0, t) = \theta_0(t)$ is satisfied, we obtain

$$\theta(x, t) = \frac{a_0}{2} + \sum_{n=1}^{\infty} \exp\left(-\sqrt{\frac{n\omega}{2D}} x\right) \left[a_n \cos\left(n\omega t - \sqrt{\frac{n\omega}{2D}} x\right) \right.$$
$$\left. + b_n \sin\left(n\omega t - \sqrt{\frac{n\omega}{2D}} x\right) \right]. \qquad (4.77)$$

The components of the sum with $n = 1$ decrease least rapidly as x increases. Thus, for large values of x, only this term and the constant term $a_0/2$ are significant, so the temperature varies approximately sinusoidally with time. ∎

4.6 Semi-infinite domains: temperature waves

Exercise 4.17

A surface is alternately exposed to one hot fluid at a temperature $+\theta_0$ for a time π/ω, then to a cold fluid at temperature $-\theta_0$ for a time π/ω, and the cycle is repeated over and over again.

(a) What is the Fourier series for the surface temperature at time t? [Hint: This was obtained for $\omega = 1$ in Block I, Chapter 3: see Example 3.2.]

(b) Show that the temperature at depth x below the surface at time t is

$$\theta(x,t) = \frac{4\theta_0}{\pi} \sum_{n=0}^{\infty} \left\{ \frac{1}{2n+1} \sin\left[(2n+1)\omega t - \sqrt{\frac{(2n+1)\omega}{2D}}\, x \right] \right.$$
$$\left. \times \exp\left[-\sqrt{\frac{(2n+1)\omega}{2D}}\, x \right] \right\}. \tag{4.78}$$

This result is illustrated in Figure 4.8, which shows the time-dependence of the temperature at the surface and at three different depths. We can again observe that as the depth increases, the amplitude of the temperature wave decreases, and that it lags further behind the surface temperature. Also, because the higher Fourier components (i.e. terms with larger n) are reduced more rapidly, the time-dependence of the temperature wave becomes essentially sinusoidal at large depths.

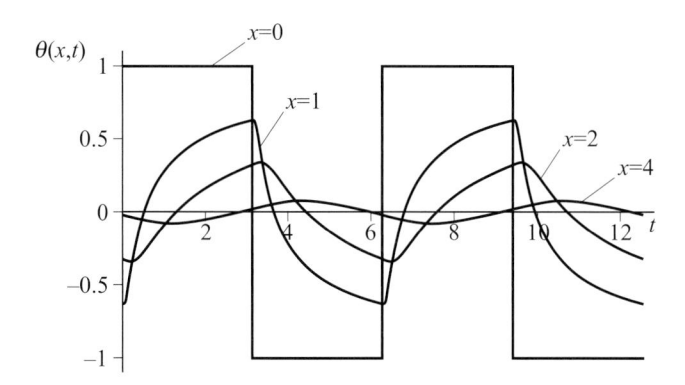

Figure 4.8 The temperature as a function of time, at depth x below a surface which is alternately heated to temperature $\theta_0 = +1$ and cooled to temperature $-\theta_0$, with a frequency $\omega = 1$. In this illustration, we have set $D = 1$.

Exercise 4.18

In desert areas, temperatures can be very high during the day and very low at night. Desert animals can escape from the extreme temperatures by burrowing into the ground.

In a desert area, the daily high and low temperatures are $40°$ C and $0°$ C respectively, and the thermal diffusivity of the dry sandy soil is 0.24×10^{-6} m^2 s^{-1}. Assuming that the flux of heat varies sinusoidally throughout the day, estimate the depth to which an animal must burrow so that the temperature of its surroundings does not exceed $25°$ C.

4.7 Outcomes

After reading this chapter, you should:
- appreciate the concept of a propagator, and be able to determine (with some guidance) the propagator for partial differential equations which are similar in structure to the diffusion equation
- be able to use a propagator to determine a temperature, concentration, or similar scalar field, by evaluating an integral
- be able to construct solutions of the diffusion equation and closely related partial differential equations using the method of separation of variables, in one or more dimensions
- be able to demonstrate orthogonality relations, and be able to use them to calculate Fourier coefficients for particular solutions of diffusion or heat flow problems
- appreciate the existence of temperature waves in periodically heated and cooled structures, and be able to calculate temperature waves in a half-space with a uniformly heated planar boundary.

4.8 Further exercises

The following exercises provide additional practice in solving typical problems of the types treated in this chapter.

Exercise 4.19

The diffusion-advection equation, describing diffusion in a moving fluid, was introduced in Chapter 3, Exercise 3.28: in one dimension it takes the form

$$\frac{\partial c}{\partial t} = -\frac{\partial}{\partial x}(vc) + D\frac{\partial^2 c}{\partial x^2},$$

where D is the diffusion constant and v is the velocity of flow of the fluid, which in this exercise is assumed to be independent of x.

(a) Can you guess the form of the propagator $K(x - x_0, t)$ for this equation?

(b) Calculate the propagator using the Fourier transform approach, as described in Section 4.2.

4.8 Further exercises

Exercise 4.20

Consider a uniform metal bar of length L which can exchange heat with its surroundings such that the temperature obeys a modified diffusion equation of the form

$$\frac{\partial \theta}{\partial t} = D\frac{\partial^2 \theta}{\partial x^2} - R(\theta - \theta_a),$$

where D is the thermal diffusivity, R is a constant (sometimes called a rate constant), and θ_a is the temperature of the surrounding air (assumed to be constant).

At time $t=0$ the bar is removed from the heat sources, so that there is no heat transferred through the ends of the bar, and the boundary conditions are

$$\frac{\partial \theta}{\partial x}(0,t) = \frac{\partial \theta}{\partial x}(L,t) = 0.$$

Using the method of separation of variables, find a general solution of the partial differential equation given above. Determine $\theta(x,t)$ in the form of a Fourier series, when the initial temperature is a specified function, $\theta_0(x)$.

Exercise 4.21

If the bar described in the previous exercise is held in contact with heat sources, so that both ends of the bar ($x=0$ and $x=L$) are at temperature θ_1, determine the steady-state temperature $\theta_0(x)$ at position x.

Exercise 4.22

A cube, with sides of length L, made from a material with thermal diffusivity D, is heated to a uniform initial temperature θ_0. At time $t=0$, it is plunged into a stream of fast-flowing liquid, so that the temperature everywhere on the surface is equal to zero for all $t > 0$. Show that the temperature at the centre of the cube is (for $t>0$)

Hard exercise

$$\theta_c(t) = \theta_0 \left(\frac{4}{\pi}\right)^3 \sum_{n_1=1}^{\infty}\sum_{n_2=1}^{\infty}\sum_{n_3=1}^{\infty} \sin\left(\frac{n_1\pi}{2}\right)\sin\left(\frac{n_2\pi}{2}\right)\sin\left(\frac{n_3\pi}{2}\right)$$
$$\times \frac{1}{n_1 n_2 n_3} \exp[-(n_1^2 + n_2^2 + n_3^2)\pi^2 Dt/L^2].$$

Solutions to Exercises in Chapter 4

Solution 4.1

Write the concentration in terms of its Fourier transform with respect to x:

$$c(x,t) = \frac{1}{\sqrt{2\pi}} \int_{-\infty}^{\infty} dk \exp(ikx)\,\tilde{c}(k,t).$$

Assuming that equation (4.4) is satisfied, we have

$$\int_{-\infty}^{\infty} dk \exp(ikx) \left[\frac{\partial \tilde{c}}{\partial t}(k,t) + Dk^2\,\tilde{c}(k,t) \right] = 0.$$

Using the fact that $(\partial^2/\partial x^2)\exp(ikx) = -k^2\exp(ikx)$, we may replace the factor of k^2 by a second derivative with respect to x, and obtain

$$\frac{1}{\sqrt{2\pi}} \int_{-\infty}^{\infty} dk \exp(ikx) \frac{\partial \tilde{c}}{\partial t}(k,t) = \frac{D}{\sqrt{2\pi}} \frac{\partial^2}{\partial x^2} \int_{-\infty}^{\infty} dk \exp(ikx)\,\tilde{c}(k,t).$$

Comparing with the first equation, we see that the left-hand side is $\partial c(x,t)/\partial t$, and the integral on the right-hand side is $\sqrt{2\pi}c(x,t)$, so $c(x,t)$ satisfies the diffusion equation.

Solution 4.2

Rearrange the given differential equation to read

$$\frac{1}{f}\frac{df}{dt} = -\alpha,$$

which can be integrated to give $\ln f = -\alpha t + C$, where $f > 0$ and C is a constant of integration. Taking the exponential of this relation gives

$$f = A\exp(-\alpha t),$$

where $A = \exp(C)$. Now apply this to equation (4.4), where $\tilde{c}$ plays the role of the dependent variable f, and the constant α is replaced by a function of k, namely $\alpha = Dk^2$. Making these substitutions gives $\tilde{c}(k,t) = A\exp(-Dk^2 t)$, where the multiplier A may be a function of k. When $t = 0$, this relation reads $\tilde{c}(k,0) = A$, which determines A, so

$$\tilde{c}(k,t) = \exp(-Dk^2 t)\,\tilde{c}(k,0).$$

Solution 4.3

(a) The initial condition for Example 4.2 may be written

$$c(x,0) = c_0 H(x-a),$$

where $H(x)$ is the step function (Heaviside function) defined in the question.

(b) In this problem, the initial condition is

$$c(x,0) = c_0[H(x+L) - H(x-L)].$$

(c) Using equations (4.7) and (4.8), we have

$$c(x,t) = \frac{c_0}{\sqrt{4\pi Dt}} \int_{-\infty}^{\infty} dx_0 \exp[-(x-x_0)^2/4Dt]\,H(x_0+L)$$

$$- \frac{c_0}{\sqrt{4\pi Dt}} \int_{-\infty}^{\infty} dx_0 \exp[-(x-x_0)^2/4Dt]\,H(x_0-L).$$

The integrals in this expression were obtained in Example 4.2, as part (a) above shows: setting a equal to $-L$ and $+L$ successively, and subtracting one expression from the other, we have

$$c(x,t) = \frac{c_0}{2}\left[\mathrm{erf}\left(\frac{x+L}{\sqrt{4Dt}}\right) - \mathrm{erf}\left(\frac{x-L}{\sqrt{4Dt}}\right)\right].$$

Solution 4.4

Again, starting from equations (4.7) and (4.8), and using the changes of variable $u = (x - x_0)/\sqrt{4Dt}$ and $v = x - x_0$, we have

$$c(x,t) = \frac{1}{\sqrt{4\pi Dt}} \int_{-\infty}^{\infty} dx_0 \exp[-(x-x_0)^2/4Dt] [A_0 + A_1 \cos(kx_0)]$$

$$= \frac{A_0}{\sqrt{\pi}} \int_{-\infty}^{\infty} du \exp(-u^2) + \frac{A_1}{\sqrt{4\pi Dt}} \int_{-\infty}^{\infty} dv \exp(-v^2/4Dt) \cos[k(x-v)].$$

Now use the identity $\cos(A - B) = \cos A \cos B + \sin A \sin B$, and evaluate the first integral:

$$c(x,t) = A_0 + \frac{A_1 \cos(kx)}{\sqrt{4\pi Dt}} \int_{-\infty}^{\infty} dv \exp(-v^2/4Dt) \cos(kv)$$

$$+ \frac{A_1 \sin(kx)}{\sqrt{4\pi Dt}} \int_{-\infty}^{\infty} dv \exp(-v^2/4Dt) \sin(kv).$$

The first of these integrals is of the form quoted in the Hint, with $\alpha = 1/4Dt$ and $\beta = k$. The second integral is zero, because the integrand is an odd function. The concentration at time t is then

$$c(x,t) = A_0 + A_1 \cos(kx) \exp(-k^2 Dt).$$

Solution 4.5

Consider the function given in equation (4.16). We can check that this satisfies the diffusion equation, by evaluating partial derivatives:

$$\frac{\partial \theta}{\partial t} = \sum_{n=0}^{\infty} -\frac{\pi^2 n^2 D A_n}{L^2} \cos\left(\frac{\pi n x}{L}\right) \exp\left(-\frac{\pi^2 n^2 Dt}{L^2}\right),$$

$$\frac{\partial \theta}{\partial x} = \sum_{n=0}^{\infty} -\frac{\pi n A_n}{L} \sin\left(\frac{\pi n x}{L}\right) \exp\left(-\frac{\pi^2 n^2 Dt}{L^2}\right),$$

$$\frac{\partial^2 \theta}{\partial x^2} = \sum_{n=0}^{\infty} -\frac{\pi^2 n^2 A_n}{L^2} \cos\left(\frac{\pi n x}{L}\right) \exp\left(-\frac{\pi^2 n^2 Dt}{L^2}\right).$$

Comparing the expressions for $\partial \theta/\partial t$ and $\partial^2 \theta/\partial x^2$, it follows that $\theta(x,t)$ satisfies the diffusion equation. We must now check that it also satisfies the boundary conditions, that is, $\partial \theta/\partial x = 0$ at both $x = 0$ and $x = L$, for all values of t. At $x = L$, $\partial \theta/\partial x = 0$ because $\sin(n\pi) = 0$ for all integer values of n. At $x = 0$, the boundary condition is satisfied because $\sin(0) = 0$.

Solution 4.6

Use equation (4.20) to obtain the Fourier coefficients for $n > 0$ via integration by parts:

$$A_n = \frac{2}{L} \int_0^L dx\, x \cos\left(\frac{n\pi x}{L}\right)$$

$$= \frac{2}{L} \left[\frac{Lx}{n\pi} \sin\left(\frac{n\pi x}{L}\right)\right]_0^L - \frac{2}{L} \frac{L}{n\pi} \int_0^L dx \sin\left(\frac{n\pi x}{L}\right)$$

$$= \frac{2L}{\pi^2 n^2} \left[\cos\left(\frac{n\pi x}{L}\right)\right]_0^L$$

$$= \frac{2L}{\pi^2 n^2} [(-1)^n - 1].$$

The Fourier coefficient for $n = 0$ is just the average value of the function, i.e.

$$A_0 = \frac{1}{L} \int_0^L dx\, x = \frac{L}{2}.$$

The Fourier coefficients are therefore

$$A_n = -\frac{2L}{\pi^2 n^2}[1 - (-1)^n] \quad \text{for } n > 0, \text{ and } \quad A_0 = \frac{L}{2}.$$

Solution 4.7

In the steady-state condition, the temperature is independent of time, so $\partial\theta/\partial t = 0$, and the temperature θ depends only upon x. The diffusion equation then reduces to the equation $d^2\theta/dx^2 = 0$, implying that $\theta(x) = \alpha x + \beta$, where α and β are constants. The mean value of the temperature is

$$\bar{\theta} = \frac{1}{L}\int_0^L dx\, \theta(x) = \tfrac{1}{2}\alpha L + \beta.$$

The temperature at position x may therefore be written

$$\theta(x) = \alpha(x - L/2) + \bar{\theta},$$

where α is the temperature gradient, and $\bar{\theta}$ is the mean temperature.

Solution 4.8

The instant after the poker is made thermally isolated, the solution of the diffusion equation in the form (4.16) is applicable. The Fourier coefficients can be deduced from those obtained in the solution to Exercise 4.6. This is because the function $\theta_0(x)$ here is obtained from that of Exercise 4.6 by multiplying by α and adding $\bar{\theta} - \tfrac{1}{2}\alpha L$. The additive constant $\bar{\theta} - \tfrac{1}{2}\alpha L$ does not contribute to the Fourier coefficients A_n for $n > 0$. The Fourier coefficients for $n > 0$ are then those of Exercise 4.6 multiplied by α, and the Fourier coefficient A_0 (which in Exercise 4.6 is $L/2$) is here replaced by the mean value of the temperature, $\bar{\theta}$. The temperature at position x and time t is then

$$\theta(x,t) = \bar{\theta} - \frac{2\alpha L}{\pi^2}\sum_{n=1}^{\infty}\frac{[1-(-1)^n]}{n^2}\cos\left(\frac{\pi n x}{L}\right)\exp\left(-\frac{\pi^2 n^2 D t}{L^2}\right).$$

Only terms corresponding to odd values of n contribute to the sum, and by setting $n = 2k+1$ we can express it as a sum which includes only these terms:

$$\theta(x,t) = \bar{\theta} - \frac{4\alpha L}{\pi^2}\sum_{k=0}^{\infty}\frac{1}{(2k+1)^2}\cos\left(\frac{(2k+1)\pi x}{L}\right)\exp\left(-\frac{(2k+1)^2\pi^2 D t}{L^2}\right).$$

Solution 4.9

Write $\theta(\boldsymbol{r},t) = f(\boldsymbol{r})g(t)$, and require that θ satisfies the diffusion equation:

$$\frac{\partial \theta}{\partial t} = f\frac{dg}{dt} = D\,\nabla^2\theta = Dg\,\nabla^2 f.$$

Dividing these equations by fg, we obtain

$$D\frac{\nabla^2 f}{f} = \frac{g'}{g}.$$

The left-hand side is independent of t, and the right-hand side is independent of $\boldsymbol{r}$. Both sides must therefore be independent of both $\boldsymbol{r}$ and t, i.e. they are equal to a constant which will be called $-\lambda D$. We therefore have the following separated equations for f and g:

$$\nabla^2 f = -\lambda f, \qquad \frac{dg}{dt} = -D\lambda g.$$

Solution 4.10

In one dimension, the Helmholtz equation is

$$\frac{d^2 f}{dx^2} + \lambda f = 0.$$

This corresponds to equation (4.13) (on setting $k^2 = \lambda$). The function $f(x) = \cos(kx)$, with $k^2 = \lambda$, is a solution of this differential equation. When $k = n\pi/L$ (with $n = 0, 1, 2, \ldots$), this choice also satisfies the Neumann boundary condition $f'(0) = f'(L) = 0$. The eigenfunctions are $f_n(x) = \cos(\pi n x/L)$, and the eigenvalues are $\lambda = k^2 = \pi^2 n^2/L^2$.

Solution 4.11

When $a = b$, the two eigenfunctions

$$f_{n,m}(x,y) = \cos(n\pi x/a)\cos(m\pi y/a), \quad f_{m,n}(x,y) = \cos(m\pi x/a)\cos(n\pi y/a)$$

(with $n \neq m$) have the same eigenvalue, namely $\lambda = \pi^2(n^2 + m^2)/a^2$, so this eigenvalue is degenerate.

Degeneracies occur whenever the lengths a and b are *rationally related*, i.e. whenever

$$\frac{a}{b} = \frac{N}{M}$$

where N and M are both integers. As an example, consider the case where $a = 3b$. The eigenfunctions

$$f_{6,1}(x,y) = \cos(6\pi x/3b)\cos(\pi y/b), \quad f_{3,2}(x,y) = \cos(3\pi x/3b)\cos(2\pi y/b)$$

both have eigenvalue

$$\lambda = \pi^2\left[\left(\frac{n}{3b}\right)^2 + \left(\frac{m}{b}\right)^2\right] = \frac{5\pi^2}{b^2}.$$

Solution 4.12

For the function $f(x) = \exp(ikx)$, we have

$$\frac{d}{dx}f(x) = ik\exp(ikx) = ikf(x), \quad \frac{d^2}{dx^2}f(x) = -k^2 f(x),$$

so f is an eigenfunction of both of the operators d/dx and d^2/dx^2.

Now apply the boundary condition $f(x + a) = f(x)$ to $f(x) = \exp(ikx)$. For any real number θ and any integer n, we have $\exp[i(\theta + 2\pi n)] = \exp(i\theta)\exp(2\pi in) = \exp(i\theta)$, so $f(x + a) = \exp[i(kx + ka)] = \exp(ikx) = f(x)$ if $ka = 2\pi n$. The boundary condition is therefore satisfied by setting $k = 2\pi n/a$. The spectrum of the operator d/dx with this boundary condition is therefore the set of eigenvalues $2\pi in/a$, where n is any integer (positive, negative or zero). Similarly, the spectrum of d^2/dx^2 is the set of eigenvalues $-(2\pi n/a)^2$, with $n = 0, \pm 1, \pm 2, \ldots$.

In the case of the operator d^2/dx^2 with the periodic boundary condition, we see that all of the eigenvalues except for $n = 0$ are doubly degenerate, because for each eigenvalue $-k^2$, the eigenfunctions $\exp(ikx)$ and $\exp(-ikx)$ are linearly independent with this same eigenvalue. There are no degeneracies for d/dx since all eigenfunctions have *distinct* eigenvalues $2\pi in/a$.

Solution 4.13

Let $f_n(\boldsymbol{r})$ be the eigenfunctions of the Helmholtz equation $(\nabla^2 + \lambda_n)f_n = 0$, satisfying the Dirichlet boundary condition $f_n(\boldsymbol{r}) = 0$ for all points $\boldsymbol{r}$ on a closed surface S enclosing a volume V.

Consider the volume integral I_{nm} defined by the first expression in the sequence of equalities

$$\begin{aligned}I_{nm} &= \int_V dV\, \boldsymbol{\nabla}\cdot(f_n\boldsymbol{\nabla}f_m - f_m\boldsymbol{\nabla}f_n) \\ &= \int_V dV\, [f_n\nabla^2 f_m - f_m\nabla^2 f_n] \\ &= (\lambda_n - \lambda_m)\int_V dV\, f_n f_m,\end{aligned}$$

where the first step uses the product rule of differentiation in the form $\boldsymbol{\nabla}\cdot(f\boldsymbol{\nabla}g) = f\nabla^2 g + \boldsymbol{\nabla}f\cdot\boldsymbol{\nabla}g$ (see equation (4.57)), and the second step uses the fact that f_n and f_m satisfy $(\nabla^2 + \lambda)f = 0$. From this expression we conclude that if $I_{nm} = 0$ and the eigenvalues λ_n and λ_m are different, then the eigenfunctions f_n and f_m are orthogonal.

Now evaluating I_{nm} using Gauss's theorem, we have

$$I_{nm} = \int_S d\boldsymbol{S}\cdot(f_n\boldsymbol{\nabla} f_m - f_m\boldsymbol{\nabla} f_n) = 0,$$

which vanishes because the Dirichlet boundary condition states that f_n and f_m are both zero everywhere on the boundary. Therefore all eigenfunctions with distinct eigenvalues are orthogonal and thus, since we assume that there are no degenerate eigenvalues, equation (4.61) holds after imposing normalisation on $f_n(\boldsymbol{r})$.

Solution 4.14

We use the same approach as in finding the solution for Neumann boundary conditions given in Subsection 4.5.2. We require a solution of $(\nabla^2 + \lambda)f = 0$ with $f(\boldsymbol{r}) = 0$ for all points $\boldsymbol{r}$ on the boundary of the same rectangle. Again, we attempt a solution by separation of variables, in the form $f(x,y) = X(x)Y(y)$. Inserting this trial solution, we again find that $X(x)$ and $Y(y)$ satisfy $X'' + k_x^2 X = 0$ and $Y'' + k_y^2 Y = 0$, where the primes denote derivatives, and the constants k_x and k_y satisfy $k_x^2 + k_y^2 = \lambda$. The boundary conditions are

$$f(0,y) = f(a,y) = 0 \quad (0 \le y \le b),$$
$$f(x,0) = f(x,b) = 0 \quad (0 \le x \le a).$$

These conditions are satisfied by finding functions $X(x)$ and $Y(y)$ which satisfy

$$X(0) = X(a) = 0, \quad Y(0) = Y(b) = 0.$$

Consider the function $X(x)$. The differential equation $X'' + k_x^2 X = 0$ is solved by $X(x) = \sin(k_x x)$. This automatically satisfies the boundary condition on $X(x)$ at $x = 0$. The boundary condition $X(a) = 0$ is satisfied by choosing k_x such that $k_x a = n\pi$, where n is an integer. Changing the sign of the integer n changes only the sign of the solution, and does not make a distinct (that is, linearly independent) eigenfunction. The value $n = 0$ is excluded because the resulting function $X(x)$ is zero everywhere. We therefore consider only the values $n = 1, 2, 3, \ldots$. Similarly, we find $Y(y) = \sin(m\pi y/a)$, with $m = 1, 2, 3, \ldots$. The eigenfunctions are therefore

$$f(x,y) = C_{n,m} \sin\left(\frac{n\pi x}{a}\right) \sin\left(\frac{m\pi y}{b}\right),$$

where the constants C_{nm} are chosen to normalise the eigenfunctions, so that

$$1 = \int_0^a dx \int_0^b dy\, [f(x,y)]^2 = C_{n,m}^2 \int_0^a dx\, \sin^2(n\pi x/a) \int_0^b dy\, \sin^2(m\pi y/b).$$

Now

$$\int_0^a dx \sin^2(n\pi x/a) = \frac{1}{2}\int_0^a dx\,[1 - \cos(2n\pi x/a)]$$
$$= \frac{a}{2} - \left[\frac{a}{4n\pi}\sin\left(\frac{2n\pi x}{a}\right)\right]_0^a = \frac{a}{2},$$

where the identity $\sin^2 x = (1 - \cos 2x)/2$ was used in the second equality. Similarly,

$$\int_0^b \sin^2(m\pi y/b)\, dy = b/2,$$

so $1 = C_{n,m}^2 \times ab/4$, and therefore $C_{n,m} = 2/\sqrt{ab}$. The eigenvalues are

$$\lambda_{nm} = k_x^2 + k_y^2 = \left(\frac{n\pi}{a}\right)^2 + \left(\frac{m\pi}{b}\right)^2, \quad n = 1,2,3,\ldots, \quad m = 1,2,3,\ldots.$$

Solution 4.15

From solution (4.75) we find

$$\frac{\partial \theta}{\partial x}(x,t) = A\sqrt{\frac{\omega}{2D}} \exp\left(-\sqrt{\frac{\omega}{2D}}\,x\right) \left[\sin\left(\phi + \omega t - \sqrt{\frac{\omega}{2D}}\,x\right) - \cos\left(\phi + \omega t - \sqrt{\frac{\omega}{2D}}\,x\right)\right],$$

so

$$\frac{\partial \theta}{\partial x}(0,t) = A\sqrt{\frac{\omega}{2D}}\,[\sin(\phi + \omega t) - \cos(\phi + \omega t)]$$

$$= A\sqrt{\frac{\omega}{2D}}\,[\sin(\omega t)(\cos\phi + \sin\phi) + \cos(\omega t)(\sin\phi - \cos\phi)].$$

In order to satisfy the boundary condition, we must have $(\partial\theta/\partial x)(0,t) = -J_0 \sin(\omega t)/\kappa$. We therefore need $\sin\phi = \cos\phi$ so that $\partial\theta/\partial x$ at $x=0$ is purely a multiple of $\sin\omega t$ (i.e. the coefficient of $\cos\omega t$ is zero). This can be satisfied by setting $\phi = 5\pi/4$, so $\sin\phi = \cos\phi = -1/\sqrt{2}$. Requiring that the resulting sinusoidal variation has the correct amplitude means that we must set

$$A\sqrt{\frac{\omega}{2D}} \times \left(-\frac{2}{\sqrt{2}}\right) = -J_0/\kappa,$$

so

$$A = \frac{J_0}{\kappa}\sqrt{\frac{D}{\omega}}.$$

Solution 4.16

Everything up to equation (4.75) remains unchanged. Applying the boundary condition, $\theta(0,t) = \theta_0 \sin(\omega t)$, we have

$$\theta(0,t) = A\cos(\phi + \omega t)$$
$$= A\cos\phi \cos\omega t - A\sin\phi \sin\omega t$$
$$= \theta_0 \sin\omega t,$$

which requires

$$A\cos\phi = 0, \quad -A\sin\phi = \theta_0,$$

which are satisfied for $\phi = -\pi/2$ and $A = \theta_0$. The required solution is therefore

$$\theta(x,t) = \theta_0 \sin\left(\omega t - \sqrt{\frac{\omega}{2D}}\,x\right)\exp\left(-\sqrt{\frac{\omega}{2D}}\,x\right).$$

Again, the solution is a travelling wave which decreases in amplitude away from the boundary.

Solution 4.17

(a) Here the function $\theta(0,t)$ is a square-wave function: $\theta(0,t) = \theta_0\,\text{sign}[\sin(\omega t)]$. The Fourier series for a square wave was determined in Block I, Chapter 3, Example 3.2. Written as a Fourier *sine* series, using equations (3.17) and (3.18) from that chapter to convert to $a_n = 0$ and $b_n = 2[1 - (-1)^n]/\pi n$, then setting $x = \omega t$ and multiplying by θ_0, we obtain

$$\theta(0,t) = \theta_0 \sum_{n=1}^{\infty} \frac{2[1-(-1)^n]}{\pi n}\sin(n\omega t).$$

(b) Each term of this series corresponds to a solution of the form determined in Exercise 4.16. Because we are solving a linear partial differential equation, we can take a linear combination of these solutions. This linear combination is chosen so that it corresponds to the Fourier series in part (a) when $x = 0$, and therefore the required solution is

$$\theta(x,t) = \frac{2\theta_0}{\pi}\sum_{n=1}^{\infty}\frac{[1-(-1)^n]}{n}\sin\left(n\omega t - \sqrt{\frac{n\omega}{2D}}\,x\right)\exp\left(-\sqrt{\frac{n\omega}{2D}}\,x\right).$$

In other words, for each term in the series, ω in Exercise 4.16 is replaced by $n\omega$, and θ_0 in Exercise 4.16 is replaced by $2\theta_0[1-(-1)^n]/\pi n$. These Fourier coefficients vanish for even values of n. Writing $n = 2k+1$, we have

$$\theta(x,t) = \frac{4\theta_0}{\pi} \sum_{k=0}^{\infty} \frac{1}{2k+1} \sin\left[(2k+1)\omega t - \sqrt{\frac{(2k+1)\omega}{2D}}\, x\right] \exp\left[-\sqrt{\frac{(2k+1)\omega}{2D}}\, x\right].$$

Solution 4.18

The amplitude of the temperature wave at the surface is $\theta_0 = \frac{1}{2}(40°\text{C} - 0°\text{C}) = 20°\text{C}$, and the average temperature is $\bar{\theta} = \frac{1}{2}(40°\text{C} + 0°\text{C}) = 20°\text{C}$ (also assumed to be the temperature reached at great depths below the surface), so the surface temperature is given by

$$\theta(0,t) = \bar{\theta} + \theta_0 \cos \omega t.$$

Hence, using Example 4.3 (with $a_0/2 = \bar{\theta}$, $a_1 = \theta_0$ and all other Fourier coefficients being zero) we have

$$\theta(x,t) = \bar{\theta} + \theta_0 \exp\left(-\sqrt{\frac{\omega}{2D}}\, x\right) \cos\left(\omega t - \sqrt{\frac{\omega}{2D}}\, x\right).$$

At the depth of the burrow, the amplitude of the temperature wave cannot exceed $25°\text{C} - 20°\text{C} = 5°\text{C}$, which is smaller than the amplitude at the surface by the ratio $R = 5°\text{C}/20°\text{C} = \frac{1}{4}$. At a depth x, the amplitude of the temperature wave is reduced by the factor $\exp(-\sqrt{\omega/2D}\,x)$. The minimum depth for the burrow, $x_{\min}$, therefore satisfies

$$R = \exp\left[-\sqrt{\frac{\omega}{2D}}\, x_{\min}\right].$$

Also, the angular frequency is $\omega = 2\pi/(24 \times 60 \times 60)\,\text{s}^{-1}$ (since the period is $24 \times 60 \times 60$, the number of seconds in 24 hours). We find (giving the final result to two significant figures)

$$x_{\min} = -\sqrt{\frac{2D}{\omega}} \ln R = \sqrt{\frac{2 \times 0.24 \times 10^{-6} \times 24 \times 60 \times 60}{2\times\pi}} \ln 4\,\text{m}$$
$$= 0.11\,\text{m} = 11\,\text{cm}.$$

Solution 4.19

(a) The effect of the fluid flow is to move particles at uniform speed v, so if $K_0(x - x_0, t)$ is the propagator for the standard diffusion equation (corresponding to the case $v = 0$), the propagator with advection included is expected to be $K(x - x_0, t) = K_0(x - x_0 - vt, t)$, that is,

$$K(x - x_0, t) = \frac{1}{\sqrt{4\pi D t}} \exp\left[-\frac{(x - x_0 - vt)^2}{4Dt}\right].$$

(b) The Fourier transform of the advection–diffusion equation (as quoted in the question) with respect to x is

$$\frac{\partial}{\partial t}\tilde{c}(k,t) = (-ikv - Dk^2)\,\tilde{c}(k,t),$$

which has the solution

$$\tilde{c}(k,t) = \exp[-(Dk^2 + ikv)t]\,\tilde{c}(k,0).$$

Following the argument presented in Section 4.2, the Fourier transform of the propagator is therefore

$$\tilde{K}(k,t) = \frac{1}{\sqrt{2\pi}} \exp(-Dtk^2) \exp(-ivtk).$$

This is the same as for the usual diffusion equation (discussed in Section 4.2), with the additional factor $\exp(-ikvt)$. From the result of Block I, Chapter 3, Example 3.6, we see that this corresponds to a translation of the inverse Fourier transform through a displacement of vt, leading to the expression for $K(x - x_0, t)$ quoted in part (a). (Alternatively, we can use the result of Exercise 3.18, which gives the inverse Fourier transform of $\exp(-ak^2)\exp(ikX)$ directly.)

Solution 4.20

Applying the method of separation of variables, writing $\theta(x,t) - \theta_a = \Delta\theta(x,t) = f(x)g(t)$ and substituting into the partial differential equation, we find that $g'f = Dgf'' - Rfg$. Dividing by fg and rearranging gives

$$R + \frac{g'}{g} = D\frac{f''}{f} = -\lambda D,$$

where $-\lambda D$ is the separation constant, which we assume should be non-positive.

Solving the equation for $f(x)$, and identifying solutions satisfying the boundary condition $(\partial/\partial x \Delta\theta)(0,t) = (\partial/\partial x \Delta\theta)(L,t) = 0$, we find normalised eigenfunctions which are identical to those obtained when $R = 0$, that is,

$$f_n(x) = \sqrt{\frac{2 - \delta_{n0}}{L}} \cos(n\pi x/L), \quad n = 0, 1, 2, 3, \ldots,$$

so $\lambda = (n\pi/L)^2$. The orthogonality relation is unchanged.

The equation for $g(t)$ is solved by $g(t) = \exp[-(\lambda D + R)t]$. The general solution of the partial differential equation is

$$\theta(x,t) = \theta_a + \sum_{n=0}^{\infty} a_n \sqrt{\frac{2 - \delta_{n0}}{L}} \cos(n\pi x/L) \exp\left[-\left(R + \frac{n^2\pi^2}{L^2}D\right)t\right].$$

Using the orthogonality relation (or equation (4.20)), the Fourier coefficients a_n are

$$a_n = \int_0^L dx f_n(x)[\theta_0(x) - \theta_a] = \sqrt{\frac{2 - \delta_{n0}}{L}} \int_0^L dx [\theta_0(x) - \theta_a]\cos(n\pi x/L).$$

Note that this solution was performed using *normalised* eigenfunctions (as described in Subsection 4.4.3). An alternative approach is to use the method set out in Section 4.3, that is, with un-normalised eigenfunctions, $\cos(n\pi x/L)$, so that a_n in both of the above equations can be replaced by $A_n\sqrt{L/(2 - \delta_{n0})}$.

Solution 4.21

Let $\Delta\theta = \theta - \theta_a$ be the difference between the temperature of the bar and the surrounding air. The steady-state solutions, satisfying $\partial\Delta\theta/\partial t = 0$, also satisfy the differential equation

$$\frac{d^2\Delta\theta}{dx^2} = \frac{R}{D}\Delta\theta,$$

which has solutions

$$\Delta\theta(x) = A\cosh\left(\sqrt{\frac{R}{D}}x\right) + B\sinh\left(\sqrt{\frac{R}{D}}x\right),$$

where A and B are constants which must be chosen to make the solution satisfy the boundary conditions. The required solution satisfies $\theta_1 = \theta(0) = \theta(L)$, so

$$\theta_1 - \theta_a = A = A\cosh\left(\sqrt{\frac{R}{D}}L\right) + B\sinh\left(\sqrt{\frac{R}{D}}L\right),$$

implying that $A = \theta_1 - \theta_a$ and

$$B\sinh\left(\sqrt{\frac{R}{D}}L\right) + (\theta_1 - \theta_a)\cosh\left(\sqrt{\frac{R}{D}}L\right) = \theta_1 - \theta_a.$$

The steady-state temperature is therefore

$$\theta_0(x) = \theta_a + (\theta_1 - \theta_a)\left[\cosh\left(\sqrt{\frac{R}{D}}x\right) + C\sinh\left(\sqrt{\frac{R}{D}}x\right)\right]$$

where
$$C = \frac{1 - \cosh(\sqrt{R/D}L)}{\sinh(\sqrt{R/D}L)}.$$

Solution 4.22

Here we must solve the Helmholtz equation $(\nabla^2 + \lambda)f = 0$ in three dimensions, for a cubic region and with Dirichlet boundary conditions. The solution is obtained by separation of variables, writing $f(x, y, z) = f_1(x)f_2(y)f_3(z)$. The approach is analogous to the treatment of the rectangular region with a Dirichlet boundary condition in two dimensions, given in Exercise 4.14. If the cube is positioned so that its sides are in the planes $x = 0$, $x = L$, $y = 0$, $y = L$, $z = 0$, $z = L$, the solution is a product of sine functions, such as $\sin(n_1\pi x/L)$, with n_1 a positive integer. The normalised eigenfunctions are

$$f_{n_1,n_2,n_3}(x,y,z) = \left(\frac{2}{L}\right)^{3/2} \sin\left(\frac{n_1\pi x}{L}\right) \sin\left(\frac{n_2\pi y}{L}\right) \sin\left(\frac{n_3\pi z}{L}\right),$$

with eigenvalues

$$\lambda = \frac{\pi^2}{L^2}(n_1^2 + n_2^2 + n_3^2).$$

The general solution is (using equation (4.53))

$$\theta(\boldsymbol{r}, t) = \sum_{\boldsymbol{n}} a_{\boldsymbol{n}} f_{\boldsymbol{n}}(\boldsymbol{r}) \exp(-\pi^2 |\boldsymbol{n}|^2 Dt/L^2),$$

where $\boldsymbol{n} = (n_1, n_2, n_3)$ and $\boldsymbol{r} = (x, y, z)$, and the $a_{\boldsymbol{n}}$ are Fourier coefficients which are chosen to make this solution match the initial temperature distribution at $t = 0$. (The summation is over all triples of positive integers $\boldsymbol{n} = (n_1, n_2, n_3)$.) The eigenfunctions are orthonormal on the cube, and the Fourier coefficients are

$$a_{\boldsymbol{n}} = \int_V dV\, \theta_0(\boldsymbol{r}) f_{\boldsymbol{n}}(\boldsymbol{r}).$$

The initial condition is that $\theta_0(\boldsymbol{r})$ is constant throughout the cube, so the Fourier coefficients are

$$a_{\boldsymbol{n}} = \theta_0 \int_V dV\, f_{\boldsymbol{n}}(\boldsymbol{r})$$
$$= \theta_0 \left(\frac{2}{L}\right)^{3/2} \int_0^L dx \sin(n_1\pi x/L) \int_0^L dy \sin(n_2\pi y/L) \int_0^L dz \sin(n_3\pi z/L)$$
$$= \theta_0 \left(\frac{2L}{\pi^2}\right)^{3/2} \frac{[1-(-1)^{n_1}][1-(-1)^{n_2}][1-(-1)^{n_3}]}{n_1 n_2 n_3},$$

using

$$\int_0^L dx\, \sin(n_1\pi x/L) = \left[\frac{-L}{n_1\pi}\cos(n_1\pi x/L)\right]_0^L = \frac{L}{n_1\pi}[1-(-1)^{n_1}]$$

and similarly for the integrals in y and z. Also, at position $\boldsymbol{r}_c = (\frac{1}{2}L, \frac{1}{2}L, \frac{1}{2}L)$, the eigenfunctions take the values

$$f_{\boldsymbol{n}}(\boldsymbol{r}_c) = (2/L)^{3/2} \sin(n_1\pi/2) \sin(n_2\pi/2) \sin(n_3\pi/2).$$

Note that $f_{\boldsymbol{n}}(\boldsymbol{r}_c) = 0$ unless n_1, n_2, n_3 are all odd, in which case the factors $[1-(-1)^{n_i}]$ are all equal to two. Inserting the Fourier coefficients and the values of $f_{\boldsymbol{n}}(\boldsymbol{r}_c)$ into the general solution gives the result quoted in the exercise.

CHAPTER 5
The central limit theorem

5.1 Introduction

Up to now this block has discussed two topics which might at first sight appear to be unrelated: Chapter 2 discussed the statistical theory of random walks, and Chapters 3 and 4 considered diffusion processes and the flow of heat. There have already been indications that random walks and diffusion are closely related: for example, the result that the second moment of a probability distribution for these processes increases linearly with time, $\langle X^2 \rangle = 2Dt$, appeared in Chapter 2 (for a situation where $D = \frac{1}{2}$, Section 2.4) and in Chapter 3 (Section 3.7). Chapter 6 will show that if a large number of particles are independently undergoing random walks, their concentration satisfies the diffusion equation. This chapter discusses a result in statistics known as the *central limit theorem* which will be used in Chapter 6 and which is very useful in its own right.

You have already seen (in Chapter 2, section 2.6) that the probability distribution for a simple random walk may be well approximated by a Gaussian function when the number of steps is large. The central limit theorem (in the form described here) is an extension of this result to the case in which the random variable has a continuous (as opposed to discrete) distribution. The theorem shows that, under quite general conditions, the probability density of a sum of N random variables approaches a Gaussian probability density function when N is large. It is a very important result in probability theory, because it helps to explain why Gaussian distributions are so commonly observed. In the context of random walks, the random variables which are summed are the displacements at each step.

Chapter 6 will consider the concentration of a large number of particles which follow independent random walks. The central limit theorem will be used there to establish that the concentration obeys the diffusion equation in systems without boundaries. A more difficult approach will also be used there to derive the diffusion equation directly in systems with boundaries.

Both this chapter and Chapter 6 will contain fewer exercises. The results which are derived are very powerful, but the derivations are difficult compared to those in most other parts of the course. If you find the derivations challenging, you should concentrate on the exercises which illustrate applications of the results. The assessment relating to this chapter will concentrate on testing your appreciation of how to apply the central limit theorem, rather than its derivation. The material in Sections 5.4 and 5.5 will not be assessed.

Some of the exercises in this chapter use proof by induction in their solution. This technique will not be required in the assessment of the course, and if you are not familiar with this approach, you should turn directly to the solutions to these exercises, where the method is explained.

This chapter relies heavily on Fourier transforms and the convolution theorem. You might find it useful to review Section 3.5 of Block I, Chapter 3, now.

5.2 The central limit theorem

Many probability density distributions occurring in practical problems are found to be very well approximated by normal distributions. In most cases this arises because the quantity of interest is the sum of a large number of random variables. It will be shown that these sums typically have a probability density which is closely approximated by a Gaussian function.

More precisely, we have the following. Consider a variable X which is the sum of N random variables x_i:

$$X = \sum_{i=1}^{N} x_i. \tag{5.1}$$

For simplicity, we assume that these variables are independent, and that they all have the same probability density, given by the function $\rho(x_i)$. Both of these assumptions can be relaxed, to a certain extent. Some of the arguments will assume certain properties of the function $\rho(x)$. For now, think of $\rho(x)$ as being a smooth function which decreases very rapidly in the limits as $x \to \pm\infty$. We shall see later that the only conditions that must be imposed on $\rho(x)$ are that the mean $\langle x \rangle$ and second moment $\langle x^2 \rangle$ both exist.

We use the symbol $\rho_X(X)$ to denote the probability density for the sum X (using the subscript X to distinguish this probability density from that of each component of the sum, $\rho(x)$). We aim to show that in the limit as $N \to \infty$, the probability density $\rho_X(X)$ of X approaches a normal distribution

$$\rho_X(X) \simeq \frac{1}{\sqrt{2\pi\sigma_X^2}} \exp[-(X - \langle X \rangle)^2 / 2\sigma_X^2]. \tag{5.2}$$

The mean $\langle X \rangle$ and variance σ_X^2 of this distribution are readily expressed in terms of the mean $\langle x \rangle$ and variance σ^2 of the x_i:

$$\langle X \rangle = N \langle x \rangle, \tag{5.3}$$

$$\sigma_X^2 = N \left\langle (x - \langle x \rangle)^2 \right\rangle = N\sigma^2. \tag{5.4}$$

The result contained in equations (5.2), (5.3) and (5.4) is known as the *central limit theorem*. The limiting probability density (5.2) is independent of the form of the probability density of each component of X (provided that the mean $\langle x \rangle$ and variance σ^2 of the probability density $\rho(x)$ exist). This insensitivity to the form of $\rho(x)$ makes the central limit theorem a very powerful result, and perhaps a rather surprising one. How can it be that the form of $\rho(x)$ becomes irrelevant as we add more variables?

As an illustration of the central limit theorem, in Figure 5.1 we display graphs of the probability densities for sums of N random variables with uniform probability density on the interval $[0, 1]$, up to $N = 5$. For $N = 1$, the probability density is a top-hat function, and for $N = 2$ you will see shortly (in Exercise 5.7) that the probability density is a function with a triangular graph. As N increases, the curves rapidly look more and more like Gaussian functions: a Gaussian probability density with the correct

5.2 The central limit theorem

mean and variance ($\frac{5}{2}$ and $\frac{5}{12}$, respectively) is plotted for comparison with the $N = 5$ case.

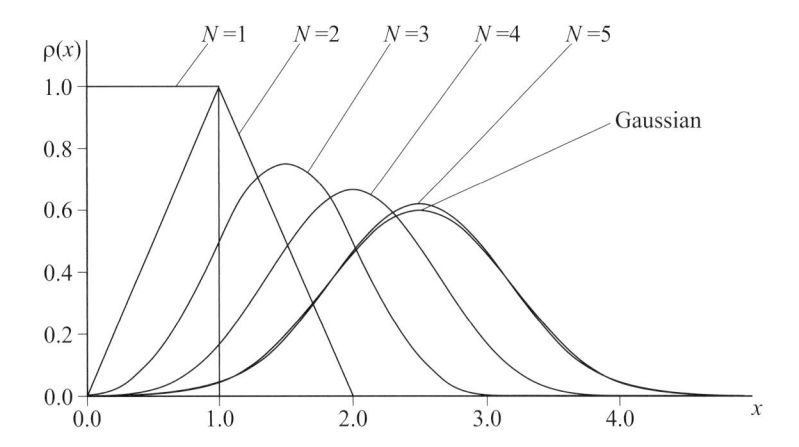

Figure 5.1 Illustrating the central limit theorem. The probability density, ρ_X, of the sum of N random variables, with each random variable having a uniform distribution, $\rho(x) = 1$, on the interval $[0, 1]$, is plotted for $N = 1, 2, 3, 4, 5$. The probability densities approach Gaussian functions as $N \to \infty$. For $N = 5$, the Gaussian approximation is already very close to the exact probability density.

> You have already seen a closely related result, for the case of a random variable which takes a discrete set of values, as opposed to the continuous case considered in this chapter. The displacement of a random walk is the sum of many independent, identically distributed steps. It was shown in Chapter 2 (Section 2.6) that this is well approximated by a normal distribution. There are slight differences: the random walk considered in Chapter 2 had a displacement which takes only integer values with a specified distribution, whereas the version of the central limit theorem stated above applies to a large class of continuous distributions. It is therefore a much more general result than that discussed in Chapter 2.

The central limit theorem is plainly a very important result, which deserves to be understood in its own right, as well as in relation to the diffusion equation. The remaining sections of this chapter will discuss results leading up to the derivation of the central limit theorem, but before tackling these sections you should consider Example 5.1 and Exercises 5.4 and 5.5, which illustrate some of its applications.

Before tackling exercises on the central limit theorem itself, you should consider the following preparatory exercises.

Exercise 5.1

Let x_1 and x_2 be two independent random variables, with mean values $\langle x_1 \rangle$ and $\langle x_2 \rangle$, respectively, and variances σ_1^2 and σ_2^2 respectively. What is the mean value of $X = x_1 + x_2$? Show that the variance of X is $\sigma_X^2 = \sigma_1^2 + \sigma_2^2$.

Exercise 5.2

Consider the sum $X_N = \sum_{i=1}^{N} x_i$ of N independent random variables x_i, all of which have the same mean $\langle x \rangle$ and variance σ^2. Show that the mean, $\langle X_N \rangle$, and variance, σ_N^2, of X_N are $\langle X_N \rangle = N \langle x \rangle$ and $\sigma_N^2 = N\sigma^2$, respectively. For convenience in this exercise, we have used a slightly different notation: X_N instead of X, and σ_N^2 in place of the more usual σ_X^2.

Hint: The second part requires proof by induction.

Exercise 5.3

Let x_i be a collection of N independent random variables, all having the same probability density, which is uniform on $[0,1]$ and zero elsewhere. What are the mean and variance of each of the x_i? What are the mean and variance of $X = \sum_{i=1}^{N} x_i$?

The following example and exercises involve knowing the probability that a Gaussian random variable is less than a certain value. This probability may be determined from the normal probability function, $N(x)$, which was defined by equation (1.55), and tabulated in Table 1.2. It is conventional to use the symbol N both for the number of variables in the sum, and for the normal distribution function. The latter will always be followed by the argument of the function in parentheses, to avoid confusion.

Example 5.1

A space satellite is assembled from 480 components. Each of the components was weighed accurately, but only the number of grams was recorded, with the part of the scale reading giving fractions of a gram being discarded (that is, the accurate weight was truncated to a whole number of grams). The manufacturer is asked to provide precise information about the weight of the completed satellite, but does not want to disrupt the assembly process to weigh everything again.

The sum of the recorded weights of the components was 19.750 kg. A clause in the contract to build the satellite will penalise the manufacturer if the weight exceeds 20 kg.

(a) By what amount might the weight of the completed satellite be expected to exceed the sum of the recorded weights of its components?

(b) Suggest a reasonable guess as to the probability distribution of the error in the weight of an individual component.

(c) What are the mean and variance of this distribution?

(d) What is the probability distribution of the amount by which the weight of the completed satellite might exceed the sum of the recorded weights of its components?

(e) Should the manufacturer be concerned about the risk that the weight will exceed 20 kg?

Solution

(a) The fractional part of the weight in grams of each component is a number between 0 and 1, about which we have no information. It is natural to model the excess weight of the component with index i as a random number x_i, with mean value $\frac{1}{2}$. The excess weight of the completed machine is then $X = \sum_{i=1}^{N} x_i$, which has mean value $\langle X \rangle = N \langle x \rangle = 480 \times 0.5\,\text{g} = 240\,\text{g}$.

The final weight of the satellite can be regarded as a random variable which has mean value $19.750 + 0.240 = 19.990\,\text{kg}$.

(b) We know that the weighing errors x_i (measured in grams) lie between 0 and 1, but have no other information about them. We therefore assume that their probability density $\rho(x)$ is a uniform distribution between 0 and 1.

(c) The mean and variance of this uniform distribution were obtained in Exercise 5.3: they are $\langle x \rangle = \frac{1}{2}$ and $\sigma^2 = \frac{1}{12}$, respectively.

5.2 The central limit theorem

(d) Using the central limit theorem and the result of Exercise 5.3, we can write down a probability density for the excess weight X in grams:

$$\rho_X(X) = \frac{1}{\sqrt{2\pi \times 480/12}} \exp\left[-\frac{(X-240)^2}{2 \times 480 \times \frac{1}{12}}\right].$$

That is, X is expected to have a normal distribution with mean value $\langle X \rangle = 240$ and standard deviation $\sigma_X = \sqrt{\frac{480}{12}} = 2\sqrt{10}$.

(e) The penalty clause will be invoked if X exceeds $20\,000 - 19\,750 = 250$, that is, 10 above the mean value. This is $10/(2\sqrt{10}) = \sqrt{10}/2 = 1.58\ldots$ multiples of the standard deviation. From Table 1.2 in Chapter 1, giving values of the normal distribution function $N(x)$, we see that the probability that a Gaussian random variable exceeds the mean by more than 1.5 standard deviations is approximately 0.067. (Note, by symmetry of the Gaussian distribution, that this probability is identical to the probability that the Gaussian random variable is *less* than the mean by more than 1.5 standard deviations, i.e. $N(-1.5)$.) There is a small but significant risk that the satellite will be overweight. ■

Exercise 5.4

The probability density for the time t taken for a light bulb to fail is given by the *Poisson distribution*, $\rho(t) = \exp(-t/\tau)/\tau$. (Of course, this applies only for $t \geq 0$, and is zero for $t < 0$.) The manufacturer claims that the mean time before failure is 100 hours. Use the central limit theorem to determine a good approximation to the probability density for the total time for ten bulbs to fail. If ten bulbs have failed after a total of 580 hours, would you doubt the manufacturer's claim?

[Hint: The times t_i for individual bulbs to fail are independent random variables, all with the same probability density. Apply the central limit theorem to the sum of the times for N bulbs to fail, $T = \sum_{i=1}^{N} t_i$, and use this to estimate the probability density for 10 bulbs failing after time T. Then determine the probability of 10 bulbs failing after 580 hours in terms of the normal probability function $N(x)$.]

Exercise 5.5

Coins can be counted by weight. It is assumed that the weights of £1 coins in a batch are independent random variables, with mean value 10 g and standard deviation 0.1 g. What are the mean and standard deviation for the weight of a batch of coins worth £100? What is the probability density of this weight?

This method of counting coins by weight is considered acceptable if the probability of counting incorrectly is less than 10^{-3}. Is this method acceptable for counting 100 £1 coins?

If this method is to be used reliably, coins from different sources should be mixed before being counted. Why might this precaution be necessary?

5.3 Distribution of sums of random variables

Before considering sums of N random variables, let us consider the probability density for the sum of just two random variables. Let X be the sum of two independent random variables, x_1 and x_2, with probability densities $\rho_1(x_1)$ and $\rho_2(x_2)$, respectively. What is the probability density $\rho_X(X)$ for $X = x_1 + x_2$?

We shall quote the result and discuss its structure before giving a derivation. The probability density for the sum $X = x_1 + x_2$ is

$$\rho_X(X) = \int_{-\infty}^{\infty} dx_1 \, \rho_1(x_1) \, \rho_2(X - x_1). \tag{5.5}$$

The form of this expression should not be surprising. We can obtain a given value of $X = x_1 + x_2$ with any possible value of x_1, but having fixed x_1 the value of x_2 is now determined: $x_2 = X - x_1$. We therefore expect to have an expression containing one integral, over x_1. We expect that this should contain the probability densities for x_1 and $x_2 = X - x_1$, hence the factors $\rho_1(x_1)$ and $\rho_2(x_2) = \rho_2(X - x_1)$.

Note that equation (5.5) indicates that ρ_X is the convolution of ρ_1 and ρ_2, as defined in Block I, Chapter 3 (Section 3.5). The probability density for the sum of two independent random variables is therefore the convolution of their individual probability densities. This may be written in symbolic form, using $\otimes$ to denote the convolution operation, as

$$\rho_X = \rho_1 \otimes \rho_2. \tag{5.6}$$

You will see that the convolution theorem will prove to be very useful in understanding the probability density for a sum of random variables.

Now we give a derivation of equation (5.5). We consider the element of probability δP that $x_1 + x_2$ lies between $X - \frac{1}{2}\delta X$ and $X + \frac{1}{2}\delta X$. Dividing δP by the width of the interval, δX, and taking the limit as $\delta X \to 0$, we obtain $\rho_X(X)$. To calculate δP, we refer to the discussion of the definition of the probability density for two random variables, discussed in Chapter 1, Subsection 1.1.5. The probability that a condition is fulfilled is obtained by integrating the probability density over the region where the condition is satisfied: see equation (1.19). In this case we have two random variables, x_1 and x_2, and the element of probability δP can be obtained by integrating their joint probability density, $\rho(x_1, x_2)$, multiplied by a function which is unity when $X - \frac{1}{2}\delta X \leq x_1 + x_2 \leq X + \frac{1}{2}\delta X$ and zero elsewhere. The region where this last function differs from zero is illustrated in Figure 5.2.

5.3 Distribution of sums of random variables

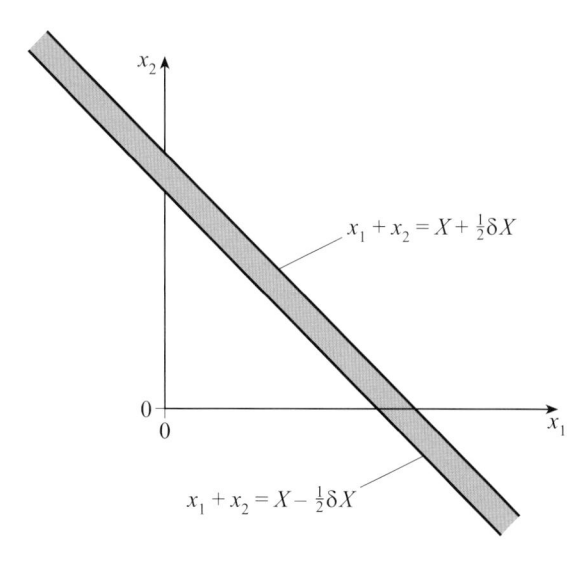

Figure 5.2 The probability that the sum $x_1 + x_2$ of two random variables lies between $X - \frac{1}{2}\delta X$ and $X + \frac{1}{2}\delta X$ is obtained by integrating their joint probability density $\rho(x_1, x_2)$ over the shaded strip (which extends to infinity in both directions)

The required function can be obtained from the 'characteristic function', introduced in Block I, Chapter 3, Section 3.3, equation 3.26. We write

$$\delta P = \int_{-\infty}^{\infty} dx_1 \int_{-\infty}^{\infty} dx_2 \, \rho(x_1, x_2) \, \chi_{\frac{1}{2}\delta X}(x_1 + x_2 - X), \tag{5.7}$$

where $\chi_\epsilon(x)$ is unity if x lies in the interval $[-\epsilon, \epsilon]$, and zero otherwise, so the factor $\chi_{\frac{1}{2}\delta X}$ selects those values of x_1 and x_2 for which the value of $X = x_1 + x_2$ lies in the required interval. The probability density of X is determined from equation (5.7) by taking a limit:

$$\rho_X(X) = \lim_{\delta X \to 0} \frac{\delta P}{\delta X}$$

$$= \lim_{\delta X \to 0} \frac{1}{\delta X} \int_{-\infty}^{\infty} dx_1 \int_{-\infty}^{\infty} dx_2 \, \rho(x_1, x_2) \, \chi_{\frac{1}{2}\delta X}(x_1 + x_2 - X). \tag{5.8}$$

This expression looks complicated, but it turns out that one of the integrals can be evaluated very easily. To see how this comes about, consider the following function $g(X)$, defined in terms of a given continuous function $f(x)$ by a combination of a limit and an integral:

$$g(X) = \lim_{\delta X \to 0} \frac{1}{\delta X} \int_{-\infty}^{\infty} dx \, f(x) \, \chi_{\frac{1}{2}\delta X}(x - X). \tag{5.9}$$

A combination of an integral and a limit of this type will be required to obtain $\rho_X(X)$ from equation (5.8). Using the fact that the characteristic function is zero outside the interval $[X - \frac{1}{2}\delta X, X + \frac{1}{2}\delta X]$, the integral is simplified:

$$g(X) = \lim_{\delta X \to 0} \frac{1}{\delta X} \int_{X - \frac{1}{2}\delta X}^{X + \frac{1}{2}\delta X} dx \, f(x). \tag{5.10}$$

Provided that $f(x)$ is continuous at $x = X$, when δX is small the function $f(x)$ is approximately equal to the constant $f(X)$ over the range of the integral in equation (5.10). This integral can therefore be approximated by $f(X)\delta X$, so $g(X) = f(X)$: that is, for any function $f(x)$ which is continuous at X, we have

$$f(X) = \lim_{\delta X \to 0} \frac{1}{\delta X} \int_{-\infty}^{\infty} dx \, \chi_{\frac{1}{2}\delta X}(x - X) \, f(x). \tag{5.11}$$

We now apply this result to the integral over x_2 in equation (5.8). We change the variable of integration x_2 to $x = x_1 + x_2$, so the function $\rho(x_1, x - x_1)$ plays the role of the function $f(x)$ in equation (5.11), and thus

$$\rho(x_1, X - x_1) = \lim_{\delta X \to 0} \frac{1}{\delta X} \int_{-\infty}^{\infty} dx_2 \, \chi_{\frac{1}{2}\delta X}(x_1 + x_2 - X) \rho(x_1, x_2). \quad (5.12)$$

Comparing equation (5.12) and (5.8), we find

$$\rho_X(X) = \int_{-\infty}^{\infty} dx_1 \, \rho(x_1, X - x_1). \quad (5.13)$$

This is a general expression for the probability density of $X = x_1 + x_2$ when x_1 and x_2 have a joint probability density $\rho(x_1, x_2)$. In the case where x_1 and x_2 are independent, the joint probability density factorises: $\rho(x_1, x_2) = \rho_1(x_1)\rho_2(x_2)$. Substituting this expression into equation (5.13), the probability density $\rho_X(X)$ is seen to be given by the convolution integral (5.5).

The following exercise shows how equations (5.5) and (5.6) can be extended to express the probability density of the sum of N identically distributed random variables in terms of the probability density of each variable.

Exercise 5.6

Let X_N be the sum of N independent and identically distributed random variables x_i, each of which has probability density ρ. Use equation (5.6) to deduce that the probability density, ρ_{X_N}, of the random variable X_N is given by repeated convolution of the function ρ with itself, i.e.

$$\rho_{X_N} = \underbrace{\rho \otimes \rho \otimes \cdots \otimes \rho}_{N-1 \text{ convolutions}}.$$

Hint: Again, this exercise requires proof by induction.

The notion of taking repeated convolution of a function may appear to be a little daunting. According to the definition introduced in Block I, Chapter 3, Section 3.5, the function $\rho_{X_2} = \rho \otimes \rho$ is obtained from $\rho(x)$ by performing a single integral. The function $\rho_{X_3} = \rho_{X_2} \otimes \rho = \rho \otimes \rho \otimes \rho$ is obtained from the function ρ_{X_2} and ρ by performing a further integration, and so on. However, you will see that the use of the convolution theorem enables us to avoid having to perform these multiple integrals.

The following example and exercise illustrate the application of equation (5.5).

Example 5.2

Let $X = x_1 + x_2$, where x_1 and x_2 are independent Gaussian random variables with mean values $\langle x_1 \rangle$, $\langle x_2 \rangle$, and variances σ_1^2, σ_2^2, respectively. Show that X has a Gaussian distribution. Determine its mean value $\langle X \rangle$ and variance σ_X^2. (This result is illustrated in Figure 5.3.)

Deduce that the sum of N independent Gaussian random variables is also a Gaussian random variable. Why is the central limit theorem a much stronger result than this?

5.3 Distribution of sums of random variables

Solution

The result can be obtained from equation (5.5) by calculating the convolution of two Gaussian functions directly. It can also be obtained by Fourier transforming, using the convolution theorem, and evaluating the inverse Fourier transform. In this case it is more convenient to follow the latter approach, which avoids doing any calculation.

Observe that each probability distribution is a Gaussian function of the form $\rho(x) = A \exp[-(x-\mu)^2/2\sigma^2]$, where μ, σ^2 are the mean and variance, and A is chosen to normalise the distribution. Now we use results from Block I, Chapter 3, in particular the result that the Fourier transform of a Gaussian is a Gaussian (Exercise 3.18). We find that the Fourier transform of each probability density is a Gaussian of the form $\tilde{\rho}(k) = a \exp(ibk) \exp(-k^2/2c^2)$, where a, b, c are constants which we need not determine. The probability density of the sum $X = x_1 + x_2$ is the convolution of the probability densities of each variable. Using the convolution theorem, the Fourier transform of the probability density ρ_X of the sum is proportional to the product of the Fourier transforms of the probability densities of each variable. We therefore have (using equation (3.51) of Block I, Chapter 3)

$$\tilde{\rho}_X(k) = \sqrt{2\pi} \times a_1 \exp(ib_1 k) \exp(-k^2/2c_1^2) \times a_2 \exp(ib_2 k) \exp(-k^2/2c_2^2)$$
$$= a_X \exp(ib_X k) \exp(-k^2/2c_X^2),$$

where a_X, b_X and c_X are three constants that can be determined in terms of a_1, b_1, c_1 and a_2, b_2, c_2. The formulae for these constants are not significant here; the important point is that the Fourier transform of ρ_X has the same form as the Fourier transform of a Gaussian function. We conclude that X is Gaussian distributed. The mean and variance are obtained from the results of Exercise 5.1, i.e. $\langle X \rangle = \langle x_1 \rangle + \langle x_2 \rangle$ and $\sigma_X^2 = \sigma_1^2 + \sigma_2^2$.

It follows by induction that the sum of N independent Gaussian random variables is Gaussian, and from Exercise 5.2 we see that the mean and variance are the sums of, respectively, the means and variances of each component. The result is therefore consistent with the central limit theorem. It is much weaker however, in the sense that the central limit theorem states that the probability density of the sum approaches a Gaussian even in cases where the elements of the sum have a non-Gaussian distribution. ■

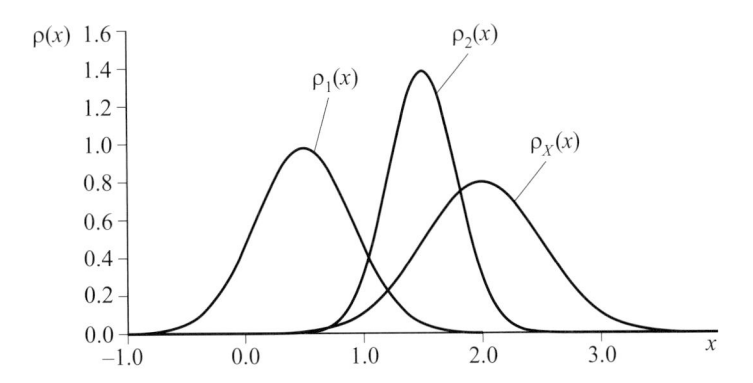

Figure 5.3 Illustrating the conclusion of Example 5.2: the probability density ρ_X of the sum of two Gaussian variables x_1 and x_2 is also a Gaussian. The parameters in this example are $\langle x_1 \rangle = \frac{1}{2}$, $\langle x_2 \rangle = \frac{3}{2}$, $\sigma_1^2 = \frac{1}{6}$, $\sigma_2^2 = \frac{1}{12}$, so $\langle X \rangle = 2$ and $\sigma_X^2 = \frac{1}{4}$.

Exercise 5.7

Two independent random variables x_1 and x_2 both take values between 0 and 1 with a uniform probability density. Show that the probability density of $X = x_1 + x_2$ is a function with a triangular graph, as illustrated in Figure 5.4.

Hint: In this case you will find it easier to calculate the convolution directly by integration, rather than through the convolution theorem. The solution of Exercise 3.22 in Block I, Chapter 3, may be useful.

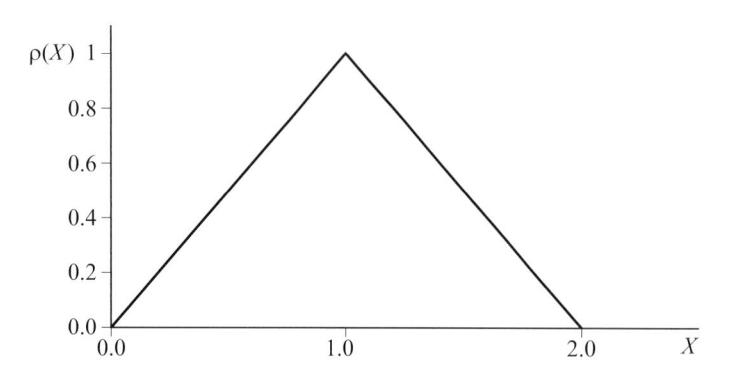

Figure 5.4 Illustrating the result of Exercise 5.7: this is the probability density ρ_X of the sum of two random variables, each of which has a uniform distribution $\rho = 1$ on the interval $[0, 1]$.

You have seen that the probability density of the sum of two independent random variables is given by the convolution $\rho_X = \rho_1 \otimes \rho_2$. If you studied the solution to Example 5.2, you will have already seen the convolution theorem used to evaluate the convolution of two Gaussian probability densities; now we shall consider the application of the convolution theorem in a more general context. Let $\tilde{\rho}_1(k)$, $\tilde{\rho}_2(k)$ and $\tilde{\rho}_X(k)$ be the Fourier transforms of the probability densities ρ_1, ρ_2 and ρ_X, respectively. Applying the convolution theorem to equation (5.6) shows that the Fourier transforms of these probability densities are related by multiplication:

You might find it useful to review the convolution theorem at this point: see Block I, Chapter 3, Section 3.5.

$$\tilde{\rho}_X(k) = \sqrt{2\pi}\,\tilde{\rho}_1(k)\,\tilde{\rho}_2(k). \tag{5.14}$$

Multiplication is a simpler operation than convolution, so this looks like a promising approach to understanding the form of the function $\tilde{\rho}_X(k)$.

If X is the sum of N independent random variables, all having the same probability density ρ, the probability density of X was shown (in Exercise 5.6) to be

For notational convenience, we now drop the subscript N from X_N in the results of Exercise 5.6.

$$\rho_X = \underbrace{\rho \otimes \rho \otimes \cdots \otimes \rho}_{N-1 \text{ convolutions}}. \tag{5.15}$$

By repeated use of the convolution theorem, the Fourier transform of ρ_X is

$$\tilde{\rho}_X(k) = (2\pi)^{(N-1)/2}[\tilde{\rho}(k)]^N. \tag{5.16}$$

It is certainly much easier to analyse repeated multiplication than repeated application of convolution integrals, and we shall now try to determine $\rho_X(X)$ by using equation (5.16) to determine $\tilde{\rho}_X(k)$, and finding the inverse Fourier transform. This suggests that we should investigate the form of the Nth power of some function, $f(x)$ say, in the limit where N is large. We do this in the next section, and find that (under very general conditions) $[f(x)]^N$ is well approximated by a Gaussian when $N \gg 1$. This result is the key to understanding the central limit theorem: we know that the Fourier transform of a Gaussian function is also a Gaussian function (see Exercise 3.18 of Block I, Chapter 3), so $\rho_X(X)$ is also expected to be well approximated by a Gaussian function.

5.4 The Gaussian approximation to $[f(x)]^N$ (Optional)

Consider a real-valued function $f(x)$, whose magnitude $|f(x)|$ has a finite global maximum at x_0 (that is, $|f(x_0)|$ is the largest value of $|f(x)|$ for any real x). We assume that the function is twice-differentiable at x_0. The first derivative of this function therefore vanishes at x_0 (that is, $f'(x_0) = 0$), so in the vicinity of x_0 it may be approximated by

$$f(x) = f_0 + \tfrac{1}{2} f_0'' \, \delta x^2 + O(\delta x^3), \tag{5.17}$$

where $\delta x = x - x_0$, $f_0 = f(x_0)$ and $f_0'' = f''(x_0)$. In the following we also assume that the second derivative is non-zero at x_0. We consider the form of the function $[f(x)]^N$, where N is a large positive integer. We aim to show that this is very well approximated by a Gaussian function (which will be obtained as equation (5.25) below). This result is illustrated in Figure 5.5, which shows $f(x)$, $[f(x)]^3$, $[f(x)]^{10}$, and Gaussian approximations $g_N(x)$ to the functions $[f(x)]^N$, for the case where $f(x)$ is a Lorentzian function

$$f(x) = \frac{1}{1 + x^2}. \tag{5.18}$$

We see that the Gaussian approximation is very good for $N = 3$, and almost indistinguishable from the function itself for $N = 10$.

The material in this section will not be assessed.

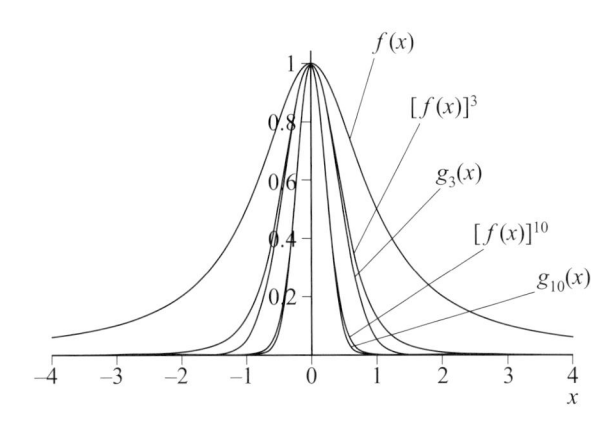

Figure 5.5 Graphs of $f(x) = 1/(1 + x^2)$, $[f(x)]^3$, $[f(x)]^{10}$, and Gaussian approximations to the latter two functions. The Gaussian approximation to $[f(x)]^N$ is $g_N(x) = \exp(-Nx^2)$.

Figure 5.6 shows similar graphs for the case where

$$f(x) = \mathrm{sinc}(x) = \begin{cases} \frac{\sin x}{x}, & x \neq 0, \\ 1, & x = 0. \end{cases} \tag{5.19}$$

Again, the Gaussian approximations are seen to be very accurate when N is large.

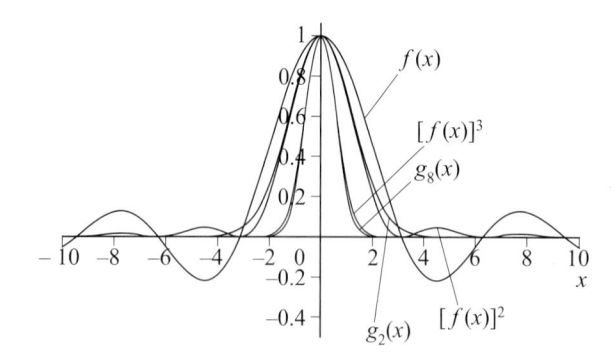

Figure 5.6 Graphs of $f(x) = \text{sinc}(x)$, $[f(x)]^2$, $[f(x)]^8$, and Gaussian approximations to the latter two functions. Here the Gaussian approximations are $g_N(x) = \exp(-Nx^2/6)$.

We now show that $[f(x)]^N$ can be approximated by a Gaussian function when N is large. It is convenient to divide $f(x)$ by its value where $|f(x)|$ is a global maximum, so that we consider a function which has a maximum value which is unity. We define functions F and F_N as follows:

$$F(x) = \frac{f(x)}{f(x_0)}, \quad F_N(x) = \left[\frac{f(x)}{f(x_0)}\right]^N. \tag{5.20}$$

By definition, $F_N(x)$ is equal to unity when $x = x_0$, and $|F_N(x)| < 1$ everywhere except at x_0. For any number a when $|a| < 1$, we have $\lim_{N \to \infty} a^N = 0$, so in the limit as $N \to \infty$, $F_N(x) \to 0$ everywhere except at $x = x_0$. When $N \gg 1$, $F_N(x)$ is approximately zero, except for a very small interval in the vicinity of the point $x = x_0$, where $F_N(x_0) = 1$. We therefore consider in detail an approximation to $F_N(x)$ which is valid in the vicinity of $x = x_0$. We write

$$F_N(x) = \exp\left(N\left[\ln F(x)\right]\right), \tag{5.21}$$

which is justified in the region of interest around the maximum of $|f(x)|$ at x_0, where $F(x)$ is positive. We now expand $F(x)$ in powers of $\delta x = x - x_0$: noting that $F'(x_0) = 0$ because x_0 is a maximum. Defining $F_0 = F(x_0)$ and $F_0'' = F''(x_0)$, we have

$$\begin{aligned} F_N(x) &= \exp\left(N \ln\left[F_0 + \tfrac{1}{2}F_0'' \delta x^2 + O(\delta x^3)\right]\right) \\ &= \exp\left(N \ln\left[1 - \tfrac{1}{2}\alpha \, \delta x^2 + O(\delta x^3)\right]\right), \end{aligned} \tag{5.22}$$

since $F_0 = 1$, and where $\alpha = -F_0''$ is a positive number because $F(x)$ has a maximum at $x = x_0$ with $f''(x_0) \neq 0$. We now use the relations

$$\begin{aligned} \ln(1+x) &= x - x^2/2 + x^3/3 + \cdots, \\ \exp(x+y) &= \exp(x)\exp(y), \\ \exp(x) &= 1 + x + x^2/2 + \cdots \end{aligned} \tag{5.23}$$

to simplify equation (5.22). First we expand the logarithm, then factor the exponential, and finally expand the exponential function containing the error term:

$$\begin{aligned} F_N(x) &= \exp\left[-\tfrac{1}{2}\alpha N \, \delta x^2 + NO(\delta x^3)\right] \\ &= \exp\left(-\tfrac{1}{2}\alpha N \, \delta x^2\right) \exp\left[NO(\delta x^3)\right] \\ &= \exp\left(-\tfrac{1}{2}\alpha N \, \delta x^2\right)\left[1 + NO(\delta x^3)\right]. \end{aligned} \tag{5.24}$$

Thus $F_N(x)$ may be approximated by a Gaussian function, with a relative error which is small when $N \, \delta x^3$ is small. The Gaussian approximation is therefore valid inside a small interval centred on x_0, of half-width $\Delta x \simeq N^{-1/3}$. At the boundaries of this interval, however, the Gaussian approximation is small, since $\exp(-\tfrac{1}{2}\alpha N \Delta x^2) \simeq \exp(-\tfrac{1}{2}\alpha N^{1/3})$ approaches

5.4 The Gaussian approximation to $[f(x)]^N$ (Optional)

zero in the limit as $N \to \infty$. Thus the Gaussian approximation (5.24) is valid inside that small interval $[x_0 - \Delta x, x_0 + \Delta x]$, whereas outside this interval both the Gaussian function and $F_N(x)$ are approximately zero. We conclude that when $N \gg 1$, the Gaussian form

$$[f(x)]^N \simeq [f(x_0)]^N \exp[-\tfrac{1}{2}\alpha N(x - x_0)^2], \tag{5.25}$$

where $\alpha = -f_0''/f_0$, is a very good approximation for all values of x.

Exercise 5.8

Calculate the Gaussian approximations for the functions used in Figures 5.5 and 5.6, and confirm that your results agree with the functions quoted in the captions.

[Hint: Because the function $\sin(x)/x$ is undefined at $x = 0$, you may find it useful to write down the Taylor series for $\sin x$ and use this to give the Taylor series of $\mathrm{sinc}(x)$. The second derivative which is required is easily extracted from this series.]

We can state our conclusions about the validity of the Gaussian approximation more formally. Consider the accuracy of the Gaussian approximation (5.24) when $|\delta x| < N^\eta$, where η is some number which we choose for our convenience. We shall choose η such that the error term in (5.24) becomes small compared to unity as $N \to \infty$. This means that $N^{1+3\eta} \ll 1$ for $N \gg 1$, so $1 + 3\eta < 0$, i.e. $\eta < -\tfrac{1}{3}$.

The Gaussian function takes the value $\exp(-\tfrac{1}{2}\alpha N^{1+2\eta})$ when $|\delta x| = N^\eta$, and this becomes small in the limit as $N \to \infty$ when $1 + 2\eta > 0$, that is, when $\eta > -\tfrac{1}{2}$. Thus we may select $-\tfrac{1}{2} < \eta < -\tfrac{1}{3}$, and find that as $N \to \infty$, the Gaussian approximation is valid even when δx is sufficiently large that the Gaussian function $\exp(-\tfrac{1}{2}\alpha N \delta x^2)$ is very small.

We shall also need a Gaussian approximation in the case where f is a complex-valued function of a real variable x, for which the global maximum of $|f(x)|$ is at x_0. In this case, the function $f(x)$ itself need not be stationary at x_0, despite the fact that $|f(x)|$ is stationary there: this point is considered in Exercise 5.9 below. Again, it is convenient to divide by $f(x_0)$, and we write $F(x) = f(x)/f(x_0)$ in the form

$$F(x) = 1 + \frac{f_0'}{f_0}\delta x + \tfrac{1}{2}\frac{f_0''}{f_0}\delta x^2 + O(\delta x^3). \tag{5.26}$$

The linear term is purely imaginary, as shown in the following exercise.

Exercise 5.9

Show that the condition that $|f|$ is stationary at x_0 implies that f'/f is purely imaginary at that point.

Hint: Write $f = R\exp(i\theta)$, where R and θ are real-valued functions of x.

Now let us consider the form of the Gaussian approximation to $F_N(x) = [F(x)]^N$ when $F(x)$ is the complex-valued function in equation (5.26). We can follow quite closely the steps of the earlier calculation for real functions. First we write $f_0' = i\theta' f_0$ (where in Exercise 5.9 we saw that θ' is real), and $f_0'' = -\alpha f_0$ (where, in general, α does not have to be real). Then equation (5.21) gives

$$F_N(x) = \exp\left[N \ln\left(1 + (f_0'/f_0)\delta x + \tfrac{1}{2}(f_0''/f_0)\delta x^2 + O(\delta x^3)\right)\right]$$
$$= \exp\left[N \ln\left(1 + i\theta'\delta x - \tfrac{1}{2}\alpha \delta x^2 + O(\delta x^3)\right)\right]. \tag{5.27}$$

Now we use the Taylor expansion of the logarithm, from equations (5.23):

$$F_N(x) = \exp\left[iN\theta'\delta x - \tfrac{1}{2}N\alpha\delta x^2 + \tfrac{1}{2}N\theta'^2\delta x^2 + NO(\delta x^3)\right]. \tag{5.28}$$

Finally, we use the results on the exponential function, also in equations (5.23), to obtain

$$F_N(x) = \exp\left(iN\theta'\delta x\right) \exp\left[-\tfrac{1}{2}N\left(\alpha - \theta'^2\right)\delta x^2\right]\left[1 + O(N\delta x^3)\right]. \quad (5.29)$$

This resembles the result for real functions, but there is an additional factor $\exp(iN\theta'\delta x)$, and α is replaced by $\alpha - \theta'^2$.

5.5 Fourier transform of a probability density (Optional)

You have seen in Section 5.3 that the probability density ρ_X for a sum of N identically distributed random variables is given by repeated convolution of the probability density for a single variable, ρ. The convolution theorem shows that the Fourier transforms of these distributions are related by equation (5.16), that is, $\tilde{\rho}_X = (2\pi)^{(N-1)/2}\tilde{\rho}^N$. In the previous section you saw that f^N approaches a Gaussian as $N \to \infty$, under quite general conditions. This suggests that $\tilde{\rho}_X$ may approach a Gaussian function. Because the inverse Fourier transform of a Gaussian is also a Gaussian function, this would lead to a justification of the central limit theorem. In order to apply this, it is necessary to characterise the properties of the Fourier transform, $\tilde{\rho}(k)$, of a probability distribution function $\rho(x)$. We shall need to establish two results concerning the Fourier transform of the probability density. First, we need to show that the Taylor expansion of $\tilde{\rho}(k)$ is related to the moments of the random variable x. Secondly, we need to show that the magnitude of the Fourier transform $\tilde{\rho}(k)$ has a global maximum at $k = 0$.

The material in this section will not be assessed.

Because the probability density is normalised, the Fourier transform

$$\tilde{\rho}(k) = \frac{1}{\sqrt{2\pi}} \int_{-\infty}^{\infty} dx \exp(-ikx)\,\rho(x) \quad (5.30)$$

takes the value $1/\sqrt{2\pi}$ at $k = 0$. By differentiating equation (5.30), we establish that derivatives of $\tilde{\rho}(k)$ at $k = 0$ are related to the moments of the probability density $\rho(x)$: for example,

The concept of normalisation of a probability density was introduced in Subsection 1.1.4.

$$\begin{aligned}\tilde{\rho}'(0) = \left.\frac{d\tilde{\rho}}{dk}\right|_{k=0} &= \frac{1}{\sqrt{2\pi}} \int_{-\infty}^{\infty} dx\,\rho(x)\,\frac{d}{dk}\exp(-ikx)\bigg|_{k=0} \\ &= \frac{-i}{\sqrt{2\pi}} \int_{-\infty}^{\infty} dx\,x\exp(-ikx)\,\rho(x)\bigg|_{k=0} \\ &= \frac{-i}{\sqrt{2\pi}} \int_{-\infty}^{\infty} dx\,x\,\rho(x) = \frac{-i}{\sqrt{2\pi}}\langle x \rangle. \end{aligned} \quad (5.31)$$

Here the notation $f(k)|_{k=0}$ means 'the function $f(k)$ evaluated at $k = 0$'.

Similarly, we find

$$\tilde{\rho}''(0) = \left.\frac{d^2\tilde{\rho}}{dk^2}\right|_{k=0} = \frac{-1}{\sqrt{2\pi}}\langle x^2 \rangle. \quad (5.32)$$

Exercise 5.10

Derive equation (5.32). Write down an expression for $\langle x^n \rangle$ (with $n \geq 0$ an integer) in terms of derivatives of $\tilde{\rho}(k)$.

It can be shown that the magnitude of $\tilde{\rho}(k)$ is greatest at $k = 0$. (This is expected because $\rho(x)$ is nowhere negative, and the oscillations of the function $\exp(ikx)$ cause cancellations when $k \neq 0$.) The following exercise provides the proof.

Exercise 5.11

Show that $|\tilde{\rho}(k)|^2$ has a global maximum at $k = 0$.

[Hint: Show that

$$|\tilde{\rho}(0)|^2 - |\tilde{\rho}(k)|^2 = \frac{1}{2\pi} \int_{-\infty}^{\infty} dx \int_{-\infty}^{\infty} dy \, [1 - \cos(k(x-y))] \, \rho(x) \rho(y), \quad (5.33)$$

and show that this integral is never negative.

Note that $|\tilde{\rho}(k)|^2 = \tilde{\rho}(k) \times [\tilde{\rho}(k)]^*$, and express both $\tilde{\rho}(k)$ and its complex conjugate as integrals involving the function $\rho(x)$.]

Note that this exercise is harder than average.

It follows that if $\tilde{\rho}(k)$ is expanded as a Taylor series about $k = 0$, the coefficients are related to the moments of $\rho(x)$:

$$\tilde{\rho}(k) = \tilde{\rho}(0) + \left.\frac{d\tilde{\rho}}{dk}\right|_{k=0} k + \frac{1}{2}\left.\frac{d^2\tilde{\rho}}{dk^2}\right|_{k=0} k^2 + \cdots$$

$$= \frac{1}{\sqrt{2\pi}} \left(1 - i\langle x \rangle k - \tfrac{1}{2}\langle x^2 \rangle k^2 + \cdots \right). \quad (5.34)$$

From the result of Exercise 5.10, we know that in general the coefficient in k^n is proportional to the moment $\langle x^n \rangle$, so

$$\tilde{\rho}(k) = \frac{1}{\sqrt{2\pi}} \left[1 - i\langle x \rangle k - \tfrac{1}{2}\langle x^2 \rangle k^2 + O(k^3) \right] \quad (5.35)$$

(provided that the moments $\langle x \rangle$ and $\langle x^2 \rangle$ exist).

This expression (5.35) is of the same form as (5.26), and we have seen that $k = 0$ is the global maximum of $|\tilde{\rho}(k)|$. We may therefore use equation (5.27) to give an approximation for $[\tilde{\rho}(k)]^N$: we see we must substitute $\delta x = k$, $f_0 = 1/\sqrt{2\pi}$, $\theta' = -\langle x \rangle$, and $\alpha = \langle x^2 \rangle$, so, from equation (5.29), we have

$$[\tilde{\rho}(k)]^N = (2\pi)^{-N/2} \exp(-iN\langle x \rangle k)$$
$$\times \exp\left[-\tfrac{1}{2} N \left(\langle x^2 \rangle - \langle x \rangle^2\right) k^2\right] \left[1 + O(Nk^3)\right]. \quad (5.36)$$

Using equation (5.16), we multiply by $(2\pi)^{(N-1)/2}$ to obtain an approximation to $\tilde{\rho}_X(k)$, and evaluate the inverse Fourier transform. Noting that most of the factors of $\sqrt{2\pi}$ cancel, we find

$$\tilde{\rho}_X(k) \simeq \frac{1}{\sqrt{2\pi}} \exp(-iN\langle x \rangle k) \exp(-N\sigma^2 k^2/2), \quad (5.37)$$

so

$$\rho_X(X) \simeq \frac{1}{2\pi} \int_{-\infty}^{\infty} dk \exp\left[ik\left(X - N\langle x \rangle\right)\right] \exp(-N\sigma^2 k^2/2)$$

$$= \frac{1}{\sqrt{2\pi N}\sigma} \exp\left[-\frac{(X - N\langle x \rangle)^2}{2N\sigma^2}\right], \quad (5.38)$$

Here we have used the Fourier transform of the Gaussian function, from Block I, Chapter 3, Exercise 3.18.

where $\sigma^2 = \langle x^2 \rangle - \langle x \rangle^2$. This is a normal distribution, with mean and variance in agreement with equations (5.3) and (5.4). We shall not discuss the error of this approximation.

We assumed that the moments $\langle x \rangle$ and $\langle x^2 \rangle$ exist. It may be possible to extend the series (5.35) to include terms of higher order in k, but in many cases the higher moments are infinite. We conclude this section by briefly considering the condition which determines whether moments exist. The crucial issue is how rapidly the function $\rho(x)$ decreases as $x \to \pm \infty$. If $\rho(x)$ decreases too slowly, the integral defining the moment will be divergent. The calculations are left to the following exercise, which is quite hard. If you are not familiar with manipulating inequalities involving integrals, you might like to turn to the solution directly.

Exercise 5.12

Consider a random variable which takes only positive values, in the case where the probability density can be bounded so that when x is larger than some constant x_0,

$$\rho(x) < \alpha x^{-\beta}. \tag{5.39}$$

Show that the moment $\langle x^n \rangle$ exists provided that $\beta > n + 1$. Conversely, show that if

$$\rho(x) > \alpha x^{-\beta} \tag{5.40}$$

when $x > x_0$, the moment $\langle x^n \rangle$ does not exist when $n > \beta - 1$.

This is another harder exercise.

Hint: The proof involves writing $\langle x^n \rangle$ as an integral, and splitting the region of integration into regions where the integrand can be bounded.

This exercise shows that continuing to expand $\tilde{\rho}(k)$ as a Taylor series may not be meaningful, because the coefficients are not well defined. However, our derivation of the central limit theorem requires only that $\langle x \rangle$ and $\langle x^2 \rangle$ exist.

5.6 Summary

This chapter started by describing the central limit theorem, and illustrating its applications. Very often a random variable is the sum of a large number of independent influences with similar magnitudes, and the central limit theorem indicates that we should expect such a random variable to have a probability density which is close to a Gaussian function. This underlies the fact that Gaussian probability densities are so commonly encountered that they are termed 'normal distributions'. We illustrated some applications of the central limit theorem, in which you are given information about the distribution of the component random variables x_i, and are asked to determine the Gaussian probability density of their sum X. The normal probability function $N(x)$, tabulated in Section 1.3, was used to make statements about the probability that X will lie inside a given interval.

We also discussed an approach to explaining how the central limit theorem arises. There are several stages to the argument, and it may be helpful to summarise these.
- We showed that the probability density for the sum of two independent random variables, $X = x_1 + x_2$, is given by the convolution of their individual probability densities: $\rho_X = \rho_1 \otimes \rho_2$.

- The convolution theorem then implies that the Fourier transforms of the probability densities are related by multiplication: $\tilde{\rho}_X = \sqrt{2\pi}\,\tilde{\rho}_1\tilde{\rho}_2$.
- We extended this to the sum of N random variables x_i all having the same probability density $\rho(x)$, the probability density of the sum $X = \sum_{i=1}^{N} x_i$ being $\rho_X(X)$. We found that the Fourier transforms of the probability densities are related by $\tilde{\rho}_X(k) = (2\pi)^{(N-1)/2}[\tilde{\rho}(k)]^N$.
- If $f(x)$ is a function with a global maximum at x_0 (and satisfying some other conditions which usually hold), we showed that in the limit as $N \to \infty$, $[f(x)]^N$ approaches a Gaussian function with its maximum at x_0.
- Combining the previous two results, we concluded that $\tilde{\rho}_X(k)$ approaches a Gaussian function as $N \to \infty$. Because the Fourier transform of a Gaussian is also a Gaussian, we concluded that $\rho_X(X)$ also approaches a Gaussian function.

At the start of the discussion we did not specify the conditions on the probability density $\rho(x)$. We found that the coefficents of k^n in the expansion of $\tilde{\rho}(k)$ are proportional to the moments $\langle x^n \rangle$. We showed that for some choices of $\rho(x)$, the moments might all be infinite after a certain value of n. However, our calculation required only that the first two terms of the Taylor series expansion of $\tilde{\rho}(k)$ exist. Thus, the central limit theorem is applicable whenever both $\langle x \rangle$ and $\langle x^2 \rangle$ exist.

5.7 Outcomes

After reading this chapter, you should:
- be aware of the scope of the central limit theorem, and be able to identify situations where it can be applied;
- be able to write down the Gaussian approximation for the probability density of the sum of independent random variables with the same mean and variance;
- be able to use tables of the normal distribution function $N(x)$ to determine the probability that a Gaussian distributed random variable lies in a given interval;
- be aware that the distribution of a sum of independent random variables is the convolution of their individual distributions, and be able to calculate these convolutions in simple cases;
- be aware that $[f(x)]^N$ approaches a Gaussian function as $N \to \infty$, under very general conditions, and of how this fact is related to the central limit theorem.

5.8 Further exercises

The following harder exercises provide an illustration of the central limit theorem for the case where the random variables which are summed have the probability density

$$\rho(x) = \begin{cases} \lambda \exp(-\lambda x), & x \geq 0, \\ 0, & x < 0, \end{cases} \qquad (5.41)$$

which describes random intervals between events. This probability density was derived in Exercise 1.32, and was also considered in Exercises 1.9, 1.16 and 5.4. It is called the *Poisson distribution*.

Exercise 5.13

Consider N independent random variables, each one of which has a Poisson distribution, with probability density given by equation (5.41). In Exercise 5.4, you found the mean and variance of this distribution: replacing τ by $1/\lambda$, these are $\langle x \rangle = 1/\lambda$ and $\langle x^2 \rangle - (\langle x \rangle)^2 = 1/\lambda^2$, respectively.

Let $\rho_{X_N}(X)$ be the probability density for the sum X_N of N such random variables. Show that this is given exactly by

$$\rho_{X_N}(X) = \begin{cases} \frac{\lambda^N X^{N-1}}{(N-1)!} \exp(-\lambda X), & X \geq 0, \\ 0, & X < 0. \end{cases} \qquad (5.42)$$

[Hint: Write $\rho_{X_{N+1}} = \rho_{X_N} \otimes \rho$, and show that if the equation above is valid for $\rho_{X_N}(X)$, then it is also valid for $\rho_{X_{N+1}}(X)$. Checking that the result is valid for $N = 1$ then implies that it is also valid for all $N > 1$. This is a further example of proof by induction.]

Exercise 5.14

Write the exact expression for the probability density $\rho_{X_N}(X)$ (valid for $X > 0$) obtained in the previous exercise in the form $\rho_{X_N}(X) = \exp[-\phi_N(X)]$. Show that the single minimum of ϕ_N is at $(N-1)/\lambda$. Using Stirling's formula (quoted in Section 2.6), show that the Taylor series expansion of $\phi_N(X)$ about its minimum may be approximated by

$$\phi_N(X) = -\ln \lambda + \tfrac{1}{2} \ln\left[2\pi(N-1)\right] + \frac{\lambda^2}{2(N-1)} \left(X - \frac{N-1}{\lambda}\right)^2. \qquad (5.43)$$

Use this result to obtain a Gaussian approximation for $\rho_{X_N}(X)$.

Exercise 5.15

Write down a Gaussian approximation to the probability density for the sum X_N of N independent random Poisson distributed variables, obtained from the central limit theorem. How does this result compare with that obtained in the previous exercise?

Calculate the exact values of $\rho_{X_N}(X)$ when $\lambda = 1$ and $N = 10$, at $X = 10$, $X = 13$ and $X = 20$, and compare with the two Gaussian approximations.

the
Solutions to Exercises in Chapter 5

Solution 5.1

In Subsection 1.2.2 it was shown that the mean value of the sum is the sum of the mean values, so $\langle X \rangle = \langle x_1 \rangle + \langle x_2 \rangle$.

The second moment of X is

$$\langle X^2 \rangle = \langle x_1^2 + 2x_1 x_2 + x_2^2 \rangle = \langle x_1^2 \rangle + \langle x_2^2 \rangle + 2\langle x_1 \rangle \langle x_2 \rangle,$$

where we have used the result that, for independent variables, $\langle x_1 x_2 \rangle = \langle x_1 \rangle \langle x_2 \rangle$; see Exercise 1.21. Recalling the relationship between the variance and the second moment given in equation (1.34), the variance of X is

$$\begin{aligned}\sigma_X^2 &= \langle X^2 \rangle - \langle X \rangle^2 \\ &= \langle x_1^2 \rangle + \langle x_2^2 \rangle + 2\langle x_1 \rangle \langle x_2 \rangle - (\langle x_1 \rangle + \langle x_2 \rangle)^2 \\ &= \langle x_1^2 \rangle - \langle x_1 \rangle^2 + \langle x_2^2 \rangle - \langle x_2 \rangle^2 \\ &= \sigma_1^2 + \sigma_2^2.\end{aligned}$$

Solution 5.2

Using equation (1.44), we find that the mean value of X_N is the sum of the mean values of each of its components:

$$\langle X_N \rangle = N \langle x \rangle.$$

The variance of X_N requires a more elaborate argument, which is an example of proof by induction. If the variance of the sum of N independent random variables is σ_N^2, then the second result derived in Exercise 5.1 shows that $\sigma_{N+1}^2 = \sigma_N^2 + \sigma^2$, because X_N and x_{N+1} are independent. Thus if the relation $\sigma_N^2 = N\sigma^2$ were true, we would have $\sigma_{N+1}^2 = (N+1)\sigma^2$. We have shown that if the relation $\sigma_N^2 = N\sigma^2$ is true for any choice of N, it is also true for $N+1$. Repeating the argument establishes that it is true for any integer greater than or equal to N. By definition $\sigma_1^2 = \sigma^2$, so the relation $\sigma_N^2 = N\sigma^2$ is true for all $N \geq 1$.

Solution 5.3

The probability density is equal to unity in the interval $[0, 1]$, and zero elsewhere (you can easily check that this is normalised.) Using equations (1.30) and (1.31), the mean and second moment of each x_i are

$$\langle x \rangle = \int_0^1 dx\, x = \tfrac{1}{2}, \qquad \langle x^2 \rangle = \int_0^1 dx\, x^2 = \tfrac{1}{3}.$$

The variance of each x_i is (using equation (1.34))

$$\sigma^2 = \langle x^2 \rangle - \langle x \rangle^2 = \tfrac{1}{3} - \left(\tfrac{1}{2}\right)^2 = \tfrac{1}{12}.$$

Using the results of Exercise 5.2, the mean and variance of the sum of N of these random variables are $\langle X \rangle = N\langle x \rangle = N/2$ and $\sigma_N^2 = N\sigma^2 = N/12$, respectively.

Solution 5.4

The probability density is $\rho(t) = \exp(-t/\tau)/\tau$ for $t > 0$, and zero for $t \leq 0$. The first and second moments are obtained using successive integration by parts:

$$\langle t \rangle = \int_0^\infty dt\, \frac{t}{\tau} \exp(-t/\tau) = \int_0^\infty dt\, \exp(-t/\tau) = \tau,$$

$$\begin{aligned}\langle t^2 \rangle &= \frac{1}{\tau} \int_0^\infty dt\, t^2 \exp(-t/\tau) \\ &= 2\tau \int_0^\infty dt\, \frac{t}{\tau} \exp(-t/\tau) \\ &= 2\tau \langle t \rangle = 2\tau^2.\end{aligned}$$

The question states that the manufacturer claims that $\langle t \rangle = 100$ hours, so we should take $\tau = 100$ hours. The variance is $\sigma^2 = \langle t^2 \rangle - \langle t \rangle^2 = \tau^2$. Assuming that $N = 10$ may be regarded as a large enough number to justify applying the central limit theorem, the probability density for the time T for N bulbs to fail is then

$$\rho_T(T) \simeq \frac{1}{\sqrt{2\pi N \tau}} \exp\left[-\frac{(T - N\tau)^2}{2N\tau^2}\right].$$

The probability for N bulbs having failed between $T = 0$ and $T = 580$ hours is

$$P(T < 580) = \int_0^{580} dT\, \rho_T(T) = N\left(\frac{580 - N\tau}{\sqrt{N}\tau}\right) - N(-\sqrt{N})$$
$$\simeq N\left(\frac{580 - N\tau}{\sqrt{N}\tau}\right), \tag{5.44}$$

where $N(x)$ is the normal probability distribution defined in Section 1.3, and the change of variable $x = (T - N\tau)/\sqrt{N}\tau$ was used. (The final approximation is justified by the fact that $N(-\sqrt{N})$ is negligibly small when N is large.) We find that

$$\frac{580 - N\tau}{\sqrt{N}\tau} = \frac{580 - 10 \times 100}{\sqrt{10} \times 100} = -1.328\ldots.$$

Referring to Table 1.2, we see that $N(-1) \simeq 0.16$ and $N(-1.5) \simeq 0.067$; $N(-1.328)$ must lie between these values. Based on the manufacturer's figure, the probability of waiting 580 hours or less for 10 bulbs to fail is therefore greater than 0.067. This is not highly improbable, and more observations might be required before disputing the manufacturer's claim.

Solution 5.5

If x is the weight of a single coin, the mean value for the weight $X = \sum_{i=1}^{N} x_i$ of $N = 100$ coins is $\langle X \rangle = N \langle x \rangle = 100 \times 10\,\text{g} = 1000\,\text{g}$. If σ^2 is the variance of the mass of an individual coin, the variance of the weight of N coins is $N\sigma^2 = 100 \times (0.1\,\text{g})^2 = 1\,\text{g}^2$, for which the corresponding standard deviation is $1\,\text{g}$. According to the central limit theorem, the probability density for the weight of 100 coins, expressed in grams, is therefore

$$\rho_X(X) = \frac{1}{\sqrt{2\pi}} \exp\left(-\frac{(X - 1000)^2}{2}\right).$$

The probability for the weight of $N = 100$ coins being closer to the mean weight of 99 or less coins than the mean weight of 100 coins is equal to the probability that the weight is less than 995 g. This is

$$P(X < 995) = \frac{1}{\sqrt{2\pi}} \int_0^{995} dX\, \exp\left(-\frac{(X - 1000)^2}{2}\right)$$
$$\simeq \frac{1}{\sqrt{2\pi}} \int_{-\infty}^{995} dX\, \exp\left(-\frac{(X - 1000)^2}{2}\right) = N(-5),$$

using the change of variable $y = X - 1000$ in the final step. The approximation involved in changing the lower limit of integration is justified by the fact that the integrand is negligibly small when $X < 0$. From Table 1.2 in Section 1.3, we see that $N(-4) \simeq 3 \times 10^{-5}$, and $N(-5)$ will be even smaller. There would be an equal probability for the weight to be greater than 1005 g. The sum of these is much less than 10^{-3}, so this method would be considered sufficiently reliable for counting batches of 100 £1 coins.

The coins should preferably be mixed before counting because batches of coins reaching the bank from some sources might be more heavily worn. This would invalidate the assumption that the weights are independent.

Solution 5.6

This is another example of proof by induction. The result was shown to be true for $N = 2$ in the derivation of equation (5.6). Assume that the result is true for the sum of N terms, and use the result (5.6) to obtain $\rho_{X_{N+1}}$:

$$\rho_{X_{N+1}} = \rho \otimes \rho_{X_N}$$
$$= \rho \otimes \underbrace{\rho \otimes \rho \otimes \cdots \otimes \rho}_{N-1 \text{ convolutions}}$$
$$= \underbrace{\rho \otimes \rho \otimes \cdots \otimes \rho}_{N \text{ convolutions}}.$$

Thus we have shown that if the result is true for X_N, it is therefore true for X_{N+1}, for any $N \geq 2$. We have already seen that the result is true for $N = 2$, so it is true in general.

Solution 5.7

The probability density of each variable is $\rho(x) = \chi(2x - 1)$, which is unity on the interval from 0 to 1, and zero elsewhere. The probability density of $X = x_1 + x_2$ is the convolution

$$\rho_X(X) = \int_{-\infty}^{\infty} dx\, \rho(X - x)\, \rho(x).$$

It follows directly from this expression, and the fact that $\rho(x)$ is one on $[0, 1]$ and zero otherwise, that the probability density ρ_X is the length of the x-interval where both $\rho(X - x)$ and $\rho(x)$ are non-zero. If $X < 0$, the integral is zero because there is no value of x for which both $\rho(x)$ and $\rho(X - x)$ are non-zero. For the same reason, when $X > 2$, the integral is zero. Now consider the case when $0 \leq X \leq 1$. In this case the x-interval where both $\rho(x)$ and $\rho(X - x)$ are non-zero is $0 \leq x \leq X$, and the length of this interval is X. Finally, consider the case when $1 \leq X \leq 2$. Here, the x-interval where both $\rho(x)$ and $\rho(X - x)$ are non-zero is $X - 1 \leq x \leq 1$, and its length is $2 - X$. Hence we can write

$$\rho_X(X) = \begin{cases} X, & \text{if } 0 \leq X \leq 1, \\ 2 - X, & \text{if } 1 < X \leq 2, \\ 0, & \text{otherwise.} \end{cases}$$

Thus the graph of ρ_X is as shown in Figure 5.4. We can also write ρ_X symbolically in other ways: for example,

$$\rho_X(X) = X\, \chi(2X - 1) + (2 - X)\, \chi(2X - 3).$$

Solution 5.8

For Figure 5.5, the function is $f(x) = 1/(1 + x^2)$. The maximum is clearly at $x_0 = 0$. We have $F(x) = f(x)$ because $f(x_0) = 1$. The Gaussian approximation, equation (5.25), is $g_N(x) = \exp(-\frac{1}{2}\alpha N x^2)$, where $\alpha = -F''(0)$.

The first two derivatives are $F'(x) = -2x/(1 + x^2)^2$ and $F''(x) = (6x^2 - 2)/(1 + x^2)^3$, so $\alpha = -F''(x_0) = 2$. The Gaussian approximation is then

$$[f(x)]^N \simeq g_N(x) = \exp(-Nx^2).$$

Alternatively, we could have noted that $(1 + x^2)^{-1} = 1 - x^2 + \cdots = 1 - \frac{1}{2}\alpha x^2 + \cdots$ (from equation (5.17)), giving $\alpha = 2$ directly.

For Figure 5.6, the function is $f(x) = \text{sinc}(x)$. Again, we have $x_0 = 0$ and $F(x) = f(x)$. In this case, it is not straightforward to calculate derivatives of $F(x)$, because $\sin(x)/x$ is undefined at $x = 0$. Instead, we use the Taylor series expansion of $\sin x$ to deduce the Taylor series for $\text{sinc}(x)$:

$$\text{sinc}(x) = \frac{1}{x}\left[x - \frac{x^3}{6} + \frac{x^5}{120} - \cdots\right] = 1 - \frac{x^2}{6} + \cdots.$$

Comparing this with the expansion $F(x) = 1 - \frac{1}{2}\alpha x^2 + \cdots$, we see that $\alpha = \frac{1}{3}$, and the Gaussian approximation is

$$[f(x)]^N \simeq g_N(x) = \exp(-Nx^2/6).$$

Solution 5.9

Write $f(x) = R(x)\exp[i\theta(x)]$, where $R \geq 0$ and θ are real-valued functions of x. Note that $|f| = R$, so $R' = 0$ at a maximum of $|f|$. We have $f' = (R' + i\theta' R)\exp(i\theta)$, so $f'/f = i\theta'$ at a maximum of $|f|$, because $R' = 0$ there. This is purely imaginary because θ is a real function.

Solution 5.10

Differentiating $\exp(-ikx)$ twice with respect to k gives $-x^2 \exp(-ikx)$. It follows that differentiating equation (5.30) twice with respect to k gives

$$\frac{d^2\tilde{\rho}}{dk^2} = \frac{-1}{\sqrt{2\pi}} \int_{-\infty}^{\infty} dx\, x^2 \exp(-ikx)\, \rho(x).$$

Setting $k = 0$ gives

$$\left.\frac{d^2\tilde{\rho}}{dk^2}\right|_{k=0} = -\frac{1}{\sqrt{2\pi}} \int_{-\infty}^{\infty} dx\, \rho(x)\, x^2 = \frac{-1}{\sqrt{2\pi}} \langle x^2 \rangle,$$

which is equation (5.32). Differentiating $\tilde{\rho}(k)$ n times with respect to k gives a factor of $(-ix)^n$ in this integral instead of $-x^2$. Multiplying both sides by $\sqrt{2\pi}\, i^n$ and setting $k = 0$ gives

$$\langle x^n \rangle = i^n \sqrt{2\pi} \left.\frac{d^n\tilde{\rho}}{dk^n}\right|_{k=0} = \sqrt{2\pi}\, i^n\, \tilde{\rho}^{(n)}(0).$$

Solution 5.11

Using equation (5.30), we have

$$|\tilde{\rho}(k)|^2 = [\tilde{\rho}(k)]^* \times \tilde{\rho}(k) = \frac{1}{2\pi} \int_{-\infty}^{\infty} dx \exp(ikx)\, \rho(x) \int_{-\infty}^{\infty} dy \exp(-iky)\, \rho(y)$$

$$= \frac{1}{2\pi} \int_{-\infty}^{\infty} dx \int_{-\infty}^{\infty} dy \exp[ik(x-y)]\, \rho(x)\, \rho(y)$$

$$= \frac{1}{2\pi} \int_{-\infty}^{\infty} dx \int_{-\infty}^{\infty} dy\, \cos[k(x-y)]\, \rho(x)\, \rho(y)$$

$$+ \frac{i}{2\pi} \int_{-\infty}^{\infty} dx \int_{-\infty}^{\infty} dy\, \sin[k(x-y)]\, \rho(x)\, \rho(y).$$

The second term must vanish, because the left-hand side is real. (It can be seen why this integral is zero by noting that the integrand has odd symmetry about the line $y = x$.) An expression for $|\tilde{\rho}(0)|^2$ is obtained by setting $k = 0$ in the above expression. Subtracting the expression for $|\tilde{\rho}(k)|^2$ from that of $|\tilde{\rho}(0)|^2$ leads to the equation quoted in the hint. Note that $\cos[k(x-y)]$ is never greater than one, therefore $1 - \cos[k(x-y)]$ can never be negative. The function $1 - \cos[k(x-y)]$ is equal to zero for all (x, y) in the case $k = 0$, but for all other values of k the product of the three non-negative functions is positive over a finite area in the (x, y)-plane. Hence $|\tilde{\rho}(0)|^2 - |\tilde{\rho}(k)|^2$ is never negative, and it follows that the maximum of $|\tilde{\rho}(k)|^2$, and hence of $|\tilde{\rho}(k)|$, must be at $k = 0$.

Solution 5.12

Consider the moment $\langle x^n \rangle$, in the case where $\rho(x) < \alpha x^{-\beta}$ for $x > x_0$:

$$\langle x^n \rangle = \int_0^\infty dx\, \rho(x)\, x^n = I_1 + I_2,$$

where the integrals I_1 and I_2 are contributions to the moment from the two regions $0 < x \leq x_0$ and $x > x_0$, respectively. Note that the lower limit of integration may be set equal to zero because the random variable is positive, so that $\rho(x) = 0$ when $x < 0$. These two integrals are bounded as follows:

$$I_1 = \int_0^{x_0} dx\, \rho(x)\, x^n < x_0^n \int_0^{x_0} dx\, \rho(x) \leq x_0^n \int_0^\infty dx\, \rho(x) = x_0^n,$$

$$I_2 = \int_{x_0}^\infty dx\, x^n \rho(x) < \int_{x_0}^\infty dx\, \alpha x^{n-\beta} = \frac{\alpha x_0^{n+1-\beta}}{\beta - n - 1}.$$

The final step used in bounding I_2 is valid provided that $n + 1 - \beta < 0$. In this case $\langle x^n \rangle$ exists, because the integrals I_1 and I_2 have been shown to be finite.

If $\rho(x) > \alpha x^{-\beta}$ for $x > x_0$, the integral I_2 satisfies

$$I_2 > \alpha \int_{x_0}^\infty dx\, x^{n-\beta},$$

and the integral on the right-hand side diverges as the upper limit approaches infinity when $n - \beta > -1$. In the case where $n > \beta - 1$, $\langle x^n \rangle$ is infinite.

Solution 5.13

Assume that the distribution for the sum of N such random variables is as quoted in the question. The probability distribution for $N + 1$ Poisson distributed random variables can then be obtained by calculating the convolution of the distribution for the sum of N Poisson random variables with the distribution for a single Poisson random variable:

$$\rho_{X_{N+1}}(X) = \int_{-\infty}^\infty dx\, \rho_{X_N}(x)\, \rho(X - x)$$

$$= \int_0^X dx\, \frac{\lambda^N x^{N-1}}{(N-1)!} \exp(-\lambda x)\, \lambda \exp[-\lambda(X - x)]$$

$$= \frac{\lambda^{N+1}}{(N-1)!} \exp(-\lambda X) \int_0^X dx\, x^{N-1}$$

$$= \frac{\lambda^{N+1} X^N}{N!} \exp(-\lambda X).$$

Note that $\rho_{X_{N+1}}(X) = 0$ for $X < 0$ because the integrand in the first line will then be zero for all x. The second line, valid provided $X \geq 0$, used the fact that the probability densities are zero for negative values, to restrict the range of integration.

Setting $N = 0$ in this expression, or $N = 1$ in the expression given in the question, we obtain the Poisson distribution itself, so the result for $\rho_{X_N}(X)$ is correct for $N = 1$. The calculation above shows that if the result is true for N, it is true for $N + 1$. The expression given in the question is therefore true for all integers $N \geq 1$.

Solution 5.14

Taking logarithms, the function $\phi_N(X)$ is

$$\phi_N(X) = -\ln[\rho_{X_N}(X)] = \lambda X - (N-1)\ln X - N\ln\lambda + \ln(N-1)!,$$

with $X > 0$. The first two derivatives with respect to X are $\phi'_N(X) = \lambda - (N-1)/X$ and $\phi''_N(X) = (N-1)/X^2$. The function $\phi_N(X)$ is stationary when $0 = \lambda - (N-1)/X$, i.e. at $X_0 = (N-1)/\lambda$. The second derivative at this point is $\phi''_N(X_0) = \lambda^2/(N-1) > 0$, so the stationary point is a minimum. The Taylor series about the minimum is

$$\phi_N(X) = \phi_N(X_0) + \tfrac{1}{2}(X-X_0)^2 \phi''_N(X_0) + O((\Delta X)^3)$$
$$= (N-1) - (N-1)\ln\left(\frac{N-1}{\lambda}\right) + \frac{\lambda^2}{2(N-1)}\left(X - \frac{N-1}{\lambda}\right)^2$$
$$\quad - N\ln\lambda + \ln(N-1)! + O((\Delta X)^3)$$
$$= -\ln\lambda + \tfrac{1}{2}\ln[2\pi(N-1)] + \frac{\lambda^2}{2(N-1)}\left(X - \frac{(N-1)}{\lambda}\right)^2$$
$$\quad + O\left((\Delta X)^3\right) + O(1/N),$$

where $\Delta X = X - (N-1)/\lambda$, and Stirling's formula for $\ln(N-1)!$ (see Chapter 2, equation (2.33)) was used to simplify the constant term.

Exponentiating gives a Gaussian approximation for the probability density:

$$\rho_{X_N}(X) \simeq \frac{\lambda}{\sqrt{2\pi(N-1)}} \exp\left[-\frac{(\lambda X - N + 1)^2}{2(N-1)}\right].$$

Solution 5.15

The mean and variance of X are $\langle X \rangle = N\langle x \rangle = N/\lambda$ and $\sigma_N^2 = N\sigma^2 = N/\lambda^2$, respectively. The central limit theorem is applicable to this problem, and gives the following approximation for $\rho_{X_N}(X)$:

$$\rho_{X_N}(X) \simeq \frac{\lambda}{\sqrt{2\pi N}} \exp\left[-\frac{(\lambda X - N)^2}{2N}\right].$$

This is slightly different from the result of the previous exercise: $N-1$ is replaced by N throughout. The difference between these expressions becomes negligible in the limit as $N \to \infty$.

In the following table, $\rho_{\text{exact}}(X)$ is given by the formula obtained in Exercise 5.13, $\rho_{N-1}(X)$ is the approximation obtained in Exercise 5.14, and $\rho_N(X)$ is the approximation obtained from the central limit theorem above.

Table 5.1

X	$\rho_{\text{exact}}(X)$	$\rho_{N-1}(X)$	$\rho_N(X)$
10	0.125	0.126	0.126
13	0.0661	0.0547	0.0804
20	0.00291	0.000160	0.000850

The Gaussian approximations are accurate close to the maximum of the probability density. Away from the maximum, their relative error is large, although the absolute error is small.

CHAPTER 6
Microscopic Derivation of the Diffusion Equation

6.1 Introduction

The material in this chapter will not be assessed, because some of it may be conceptually difficult, particularly if you had little contact with the concepts of probability and statistics before starting the course. It will be possible to gain full marks without having read this chapter.

This chapter is included because it finally draws together the two strands discussed earlier, namely the macroscopic and deterministic description of diffusion considered in Chapters 3 and 4, and the microscopic description in terms of the random walk model for the motion of molecules, which was introduced in Chapter 2, and supported by the discussion of probability and statistics in Chapters 1 and 5. Studying this chapter will deepen your understanding of random walks and diffusion, and we recommend that you read it.

You have seen that there are close connections between diffusion and random walks. Let us start by recalling some of the similarities. In Section 3.7 you saw that the diffusion equation

$$\frac{\partial c}{\partial t} = D \frac{\partial^2 c}{\partial x^2} \tag{6.1}$$

has a solution in the form of a Gaussian function

$$c(x, t) = \frac{N}{\sqrt{4\pi D t}} \exp\left(-\frac{x^2}{4Dt}\right) \tag{6.2}$$

which represents (for $t > 0$) the concentration coming from N particles, all of which are initially located at position $x = 0$ at time $t = 0$. As discussed in Section 3.7, we can obtain the probability density for the position of a single particle by dividing the concentration by the number of particles: $\rho(x, t) = c(x, t)/N$. The variance of this Gaussian probability density is proportional to the time elapsed since the start of the diffusion process:

$$\langle x^2 \rangle = \int_{-\infty}^{\infty} dx \, x^2 \rho(x, t) = 2Dt. \tag{6.3}$$

This result was discussed in Section 3.7.

Similar expressions occurred in our study of random walks in Chapter 2, and these indicate that the random walk is the microscopic process which causes diffusion. In Sections 2.5 and 2.7, we discussed the probability for a random walk which makes steps $+\delta X$ or $-\delta X$ each with probability $\frac{1}{2}$, at a sequence of times separated by δT. The probability $P(X,T)$ to reach position X at time T satisfies

$$P(X, T + \delta T) = \tfrac{1}{2}[P(X - \delta X, T) + P(X + \delta X, T). \tag{6.4}$$

In Section 2.7 it was argued that this equation may be thought of as a discrete form of the diffusion equation, with diffusion constant $D = \delta X^2 / 2\, \delta T$. Section 2.6 considered an approximate solution of equation (6.4), in the case where $\delta X = \delta T = 1$ (so that $D = \tfrac{1}{2}$). It was shown that the solution of equation (6.4), starting with the particle located at $X = 0$ when $T = 0$, is well approximated by a Gaussian function

$$P_{\mathrm{app}}(X, T) = \frac{2}{\sqrt{2\pi T}} \exp(-X^2/2T), \tag{6.5}$$

which is of the same form as equation (6.2) if we set $D = \tfrac{1}{2}$ (apart from the multiplying constant). The variance of the probability distribution for the simple random walk was shown (Section 2.4) to be

$$\langle X^2 \rangle = T \tag{6.6}$$

when $\delta X = \delta T = 1$. Again, this is consistent with the result obtained from the diffusion equation, namely equation (6.3), when $D = \tfrac{1}{2}$.

The connections between random walks and diffusion are clearly very close, because they both have a Gaussian solution with a variance that is proportional to time. It has already been explained that diffusion results from the random motion of molecules, but the derivation of the diffusion equation in Chapter 3 did not start from this microscopic viewpoint. In Chapter 3 (Section 3.5) the diffusion equation was derived from Fick's law, relating the flux density to the concentration gradient, $\boldsymbol{J} = -D\boldsymbol{\nabla} c$. This is an intuitively appealing assumption, but it was not justified from any microscopic model of the particle motion. In this chapter, we shall obtain the diffusion equation starting from the assumption that the diffusing molecules follow independent random walks, in which the particles make a very large number of very short random steps, as illustrated in Figure 2.4.

The first task, addressed in Section 6.2, is to consider a generalisation of the random walk model which is suitable for modelling the motion of molecules. Section 6.3 discusses a derivation of the diffusion equation which is based upon the central limit theorem considered in Chapter 5. This derivation is applicable only for an infinite medium (that is, a region without boundaries), and it is desirable to have a more widely applicable derivation. Section 6.4 discusses a further generalisation of the random walk, and derives a general equation for its probability density. This equation is called the *generalised diffusion equation* or the *Fokker–Planck equation*. The diffusion equation is a special case of the Fokker–Planck equation.

6.2 Continuous random walks

In Chapter 2 we introduced the idea that a molecule moving in a gas undergoes many collisions with other molecules, illustrated schematically in Figure 2.4. The motion of a molecule can be modelled by a random walk, but we must extend the definition of a random walk in various ways before it can be used to model the motion of molecules.

First, the collisions between molecules in a gas or liquid are very frequent, so the time δt between collisions is very short. In practice, the time between collisions is so short (typically 10^{-10} s, as we shall see in Exercise 6.2) that we can take the limit as $\delta t \to 0$. In this limit, the displacement $x(t)$ of the particle becomes a continuous function of the position. We describe the limit as $\delta t \to 0$ in Subsection 6.2.1. Another minor extension of the earlier models for random walks which is introduced there is that the displacement at each step is a continuous, rather than discrete, random variable.

The other generalisation of the random walk that is required is its extension to three dimensions: this is considered in Subsection 6.2.2.

6.2.1 The continuous random walk in one dimension

In Chapter 2 (Sections 2.3 and 2.4) we investigated a discrete random walk, with the displacement X allowed to take only integer values. The displacements were changed at times T, with time having unit spacing. Here we consider a random walk where the displacement Δx_n at the nth step can take a continuous range of values, with probability density $\rho_{\rm s}(\Delta x)$. The steps are separated by a short time interval δt, so the displacement after time t is

$$x(t) = \sum_{n=1}^{M} \Delta x_n, \quad \text{where} \quad M = \text{Int}(t/\delta t). \tag{6.7}$$

Note that the subscript 's' in $\rho_{\rm s}$ stands for 'step'.

Here Int(x) means 'integer part of x'.

It is assumed that the displacements Δx_n are independent random variables. This expression is analogous to equation (2.13) of Chapter 2. Initially, we shall consider the case where the mean value of Δx_n is zero, and where the variance of Δx_n is independent of the current position of the particle and of time, so

$$\langle \Delta x_n \rangle = 0 \quad \text{and} \quad \langle \Delta x_{n_1} \Delta x_{n_2} \rangle = \sigma^2 \delta_{n_1 n_2} \tag{6.8}$$

Here $\delta_{n_1 n_2}$ is the Kronecker delta symbol.

for some constant σ. These relations are analogous to equations (2.3) and (2.4) of Chapter 2; here σ is the typical size of the displacement at each step. Figure 6.1 illustrates several realisations of the random walk described by equations (6.7) and (6.8).

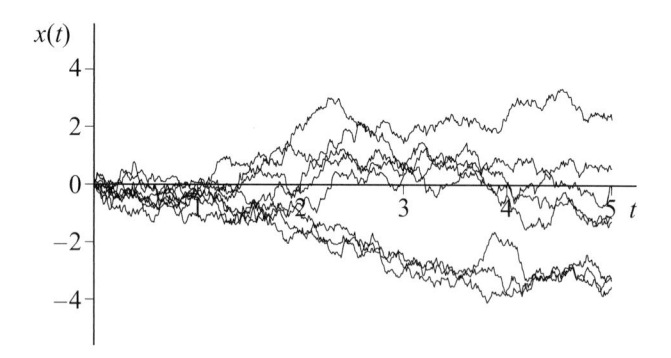

Figure 6.1 Eight realisations of the random walk described by equations (6.7) and (6.8). Here $\sigma = 0.1$, $\delta t = 0.01$, and ρ_s is a Gaussian probability density.

Following the approach introduced in Chapter 2 (Section 2.4), we describe the properties of this random walk by calculating the mean and variance of the displacement after time t. Taking the mean value of equation (6.7), and using equations (1.44) and (6.8), we see immediately that

$$\langle x(t) \rangle = \sum_{n=1}^{M} \langle \Delta x_n \rangle = 0. \tag{6.9}$$

This reflects the fact that the displacement $x(t)$ is equally likely to be positive or negative. The typical size of the displacement is understood by calculating the mean of $[x(t)]^2$, using equation (6.8):

$$\begin{aligned}
\langle [x(t)]^2 \rangle &= \left\langle \left[\sum_{n=1}^{M} \Delta x_n \right]^2 \right\rangle \\
&= \left\langle \sum_{n_1=1}^{M} \sum_{n_2=1}^{M} \Delta x_{n_1} \Delta x_{n_2} \right\rangle \\
&= \sum_{n_1=1}^{M} \sum_{n_2=1}^{M} \langle \Delta x_{n_1} \Delta x_{n_2} \rangle \\
&= \sum_{n_1=1}^{M} \sum_{n_2=1}^{M} \sigma^2 \delta_{n_1 n_2} \\
&= \sigma^2 \sum_{n_1=1}^{M} 1 = M\sigma^2 \simeq \frac{\sigma^2 t}{\delta t}.
\end{aligned} \tag{6.10}$$

Because $M = \text{Int}(t/\delta t)$, the error in the final approximation is no greater than σ^2.

We want to use this model to describe the motion of molecules, where the time between collisions is very short. We therefore consider random walks for which the time step δt is very small, and we shall take the limit as $\delta t \to 0$, so that the position $x(t)$ is a continuous function exhibiting a random walk. We want this continuous random walk to describe motion with a given value of the diffusion constant D, so we should choose the values of σ and δt such that $\langle [x(t)]^2 \rangle = 2Dt$. We do this by writing

$$\sigma^2 = 2D\,\delta t, \tag{6.11}$$

so equation (6.10) becomes equivalent to equation (6.3). Because the error in the final step of equation (6.10) is less than σ^2, and σ^2 is proportional

6.2 Continuous random walks

to δt, the error in the final approximate step in equation (6.10) vanishes as $\delta t \to 0$.

> We emphasise that this is a simplified model for the motion of molecules, which can be improved upon in various ways (for example, by taking account of the fact that the interval between collisions may not be constant). Our objective here is to show how the diffusion equation arises from a microscopic model of particle motion, and for this purpose our simple model is sufficient. More realistic and detailed models confirm that the diffusion equation describes the variation of concentration, and also enable the diffusion constant to be calculated from first principles.

In Section 2.5 we considered the probability $P(X, T)$ for the displacement of a discrete random walk, and in Section 2.6 it was shown that this is well approximated by a Gaussian function. Now let us consider the analogous question for our continuous random walk. The displacement $x(t)$ for the continuous random walk can take a continuous range of values, so its distribution must be described by a probability density, which we denote $\rho(x,t)$. In the limit as $\delta t \to 0$, the number of elements in the sum (6.7) approaches infinity, so the central limit theorem may be applicable. The other conditions of the central limit theorem as stated in Chapter 5 are met: we have a sum of independent and identically distributed random variables Δx_n, for which the mean value $\langle \Delta x_n \rangle = 0$ and variance $\langle \Delta x_n^2 \rangle = 2D\,\delta t$ both exist. The mean and variance of $x(t)$ were obtained above: we have $\langle x(t) \rangle = 0$ from equation (6.9) and $\langle x(t)^2 \rangle = 2Dt$ from equations (6.10) and (6.11). Applying the central limit theorem (using equations (5.2) to (5.4)), we see that in the limit as $\delta t \to 0$, the probability density for the displacement $x(t)$ is a Gaussian function of x:

$$\rho(x,t) = \frac{1}{\sqrt{4\pi D t}} \exp(-x^2/4Dt). \tag{6.12}$$

This is identical in form to equation (6.2) (after dividing by N, so that the solution represents a probability density for a diffusing particle). That equation was previously shown (Exercise 3.21) to satisfy the diffusion equation. Solutions of the form (6.12) were plotted in Figure 3.10.

6.2.2 Random walks in two and three dimensions

The motion of molecules diffusing in a gas or liquid occurs in three dimensions. (Figure 2.4 is a two-dimensional schematic illustration of this motion.) As well as considering the limiting case where the random steps are extremely short, we must also consider how to describe a random walk in three dimensions. In this section we state the equations that define the random walk in three dimensions. This random walk is used as a model for the microscopic motion of particles undergoing diffusion. In Sections 6.3 and 6.4 it is shown that this model leads to a derivation of the diffusion equation.

In its simplest form (considered in this section), the extension of the definition of a random walk to three dimensions is straightforward: each Cartesian component of the position of a particle makes an independent random walk. Stating this more formally, we can describe a continuous random walk in three dimensions as follows. The displacements at the nth time step in the directions of the x-, y- and z-axes, Δx_n, Δy_n and Δz_n, are independent random variables, satisfying the following equations:

$$\begin{gathered} \langle \Delta x_n \rangle = \langle \Delta y_n \rangle = \langle \Delta z_n \rangle = 0 \\ \langle \Delta x_n^2 \rangle = \langle \Delta y_n^2 \rangle = \langle \Delta z_n^2 \rangle = 2D\,\delta t \\ \langle \Delta x_n \Delta y_n \rangle = \langle \Delta x_n \Delta z_n \rangle = \langle \Delta y_n \Delta z_n \rangle = 0. \end{gathered} \tag{6.13}$$

The first line of the above equations specifies that the mean values of the displacements in the three directions are zero. The second line specifies that the mean-squared displacements in each direction are all equal to $2D\,\delta t$, as for the one-dimensional case. The third line states that the displacements in different directions are all uncorrelated with each other (which is a consequence of the fact that they are independent). As in Subsection 6.2.1, we take the limit as $\delta t \to 0$ in order to model the very frequent collisions experienced by a molecule undergoing diffusion.

There is an alternative notation which is more compact than equations (6.13), and which will be used in preference later in this chapter. Let the position of the particle be $\boldsymbol{r}(t)$, where $\boldsymbol{r}$ has components $(x_1, x_2, x_3) = (x, y, z)$. The vector displacement at time $n\,\delta t$ is $\Delta \boldsymbol{r}_n$. The components can be labelled by an index i, so that the displacement may be written

$$\Delta \boldsymbol{r}_n = (\Delta x_n, \Delta y_n, \Delta z_n) = (\Delta x_{1,n}, \Delta x_{2,n}, \Delta x_{3,n}). \tag{6.14}$$

With this notation, equations (6.13) can be expressed more concisely as follows. The mean displacements are equal to zero,

$$\langle \Delta x_{i,n} \rangle = 0, \quad i = 1, 2, 3, \tag{6.15}$$

and the mean values of products of displacements are

$$\langle \Delta x_{i,n} \Delta x_{j,m} \rangle = 2\delta_{ij}\delta_{nm} D\,\delta t, \quad i, j = 1, 2, 3. \tag{6.16}$$

The factor δ_{nm} is the Kronecker delta symbol, indicating that for *different time steps* ($n \neq m$), the displacements $\Delta x_{i,n}$ have no correlation. Figure 6.2 is a schematic illustration of the motion described by equations (6.15) and (6.16) in two dimensions.

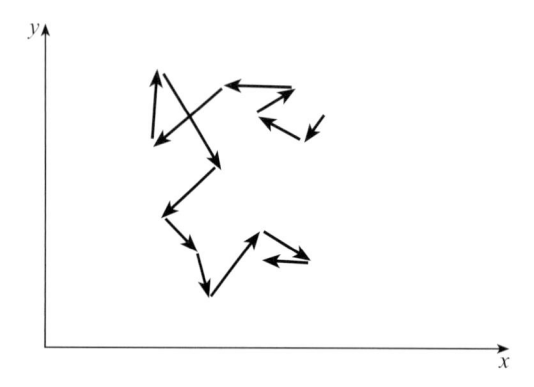

Figure 6.2 Schematic illustration of the steps of a random walk in two dimensions

Exercise 6.1

Taking the limit as $\delta t \to 0$ such that the diffusion constant D remains fixed, write down a probability density for the particle to reach (x, y, z) after time t, starting at the origin when $t = 0$.

How does this compare with the result of Exercises 3.2 and 3.24 in the two-dimensional case?

Hint: Recall that probability densities for independent variables are multiplied.

Exercise 6.2

The diffusion constant D for carbon dioxide molecules diffusing through air is measured to be $1.29 \times 10^{-5}\,\text{m}^2\,\text{s}^{-1}$. Between collisions, the molecules move at a speed v which is comparable to the speed of sound, roughly $260\,\text{m}\,\text{s}^{-1}$. Make rough estimates of the typical distance that the carbon dioxide molecules travel between collisions, and of the number of collisions each carbon dioxide molecule experiences every second. (Because you are asked for rough estimates, it is sufficient to use formulae from Subsection 6.2.1 which apply to a one-dimensional model.)

Hint: The typical distance between collisions is $\sigma \simeq v\,\delta t$. The values of σ and δt are related to the diffusion constant D through equation (6.11).

6.3 From random walks to the diffusion equation

You have now seen how to extend that definition of the random walk to three dimensions, and to the limit where the time step, δt, is taken to zero. Now it will be shown how this model for the microscopic motion of molecules leads to the diffusion equation. In this section you will be shown a simple derivation which uses the central limit theorem, but which applies only when the particles move in a region without boundaries. Section 6.4 will give a much more general derivation.

In the following, we assume that individual particles follow independent random walks, similar to that shown schematically in Figure 2.4, and show that the concentration of particles satisfies the diffusion equation. For simplicity, we consider a situation in which the concentration depends upon only one Cartesian coordinate, so that we need to consider only a one-dimensional situation; the extension to three dimensions is straightforward. In this section we assume that particles are diffusing in an infinite medium; the more difficult case of a medium with boundaries is treated in the next section.

Consider the concentration of particles that results from N particles being released at position x_0 at time t_0. What is the concentration of particles at time t? The probability for any given particle moving into the small interval between x and $x + \delta x$ is $\delta P = \rho(x - x_0, t - t_0)\,\delta x$, where $\rho(\Delta x, \Delta t)$ is the probability density for moving a distance Δx after time Δt. If N is very large, it follows (from the definition of probability – see Section 1.1) that the number of particles in this interval is approximately

$$\delta N = N\delta P = N\rho(x - x_0, t - t_0)\,\delta x. \tag{6.17}$$

Using equation (3.5), the concentration in this interval is

$$c(x, t) = \frac{\delta N}{\delta x} = N\rho(x - x_0, t - t_0). \tag{6.18}$$

For a particle executing a random walk, we have already seen that the probability density ρ is the Gaussian function of equation (6.12). In the case where the particles are all initially positioned at x_0 at time t_0, the concentration is (for $t > t_0$)

$$c(x, t) = \frac{N}{\sqrt{4\pi D(t - t_0)}} \exp\left[-\frac{(x - x_0)^2}{4D(t - t_0)}\right]. \tag{6.19}$$

This expression is in exact agreement with a solution of the diffusion equation, namely equation (3.57). This establishes that the concentration of particles undergoing independent random walks obeys the diffusion equation, for the case where all of the particles start at the same position, and diffuse in an infinite medium.

If the particles do not all start at the same position, the argument can easily be generalised. Let us assume that the N particles have initial concentration $c(x, t_0)$ at time t_0. The number of particles initially in the short interval from x_0 to $x_0 + \delta x_0$ is $c(x_0, t_0)\, \delta x_0$. The number of these particles which reach the short interval from x to $x + \delta x$ at time t (with $t > t_0$) is given by substituting $c(x_0, t_0)\, \delta x_0$ for N in equation (6.17): this number is

$$\delta \mathcal{N}(x_0) = \rho(x - x_0, t - t_0)\, \delta x \times c(x_0, t_0)\, \delta x_0. \tag{6.20}$$

To determine the total number of particles in the interval from x to $x + \delta x$ at time t, we must sum contributions of the form (6.20) coming from a set of short intervals covering the whole line (see Figure 6.3). If we take the nth interval to be $[n\, \delta x_0, (n+1)\, \delta x_0]$, the number of particles in the interval of width δx at position x is

$$\begin{aligned}
\delta N &= \sum_{n=-\infty}^{\infty} \delta \mathcal{N}(n\, \delta x_0) \\
&= \sum_{n=-\infty}^{\infty} \rho(x - n\, \delta x_0, t - t_0)\, \delta x \times c(n\, \delta x_0, t_0)\, \delta x_0.
\end{aligned} \tag{6.21}$$

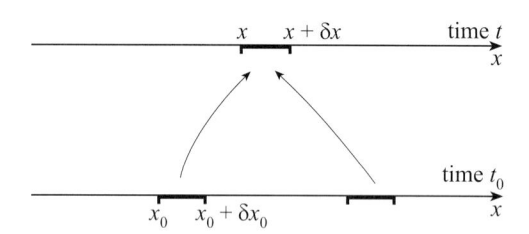

Figure 6.3 The number of particles in the interval $[x, x + \delta x]$ at time t is the sum of contributions from all of the intervals $[x_0, x_0 + \delta x_0]$ at time t_0

In the limit as $\delta x_0 \to 0$, this becomes an integral

$$\delta N = \delta x \int_{-\infty}^{\infty} dx_0\, \rho(x - x_0, t - t_0)\, c(x_0, t_0). \tag{6.22}$$

To determine the concentration at x, we divide both sides by δx as before. Now we use the Gaussian form for $\rho(x - x_0, t - t_0)$ (equation (6.12)) obtained from the central limit theorem. We find

$$c(x, t) = \frac{1}{\sqrt{4\pi D(t - t_0)}} \int_{-\infty}^{\infty} dx_0\, \exp\left[-\frac{(x - x_0)^2}{4D(t - t_0)}\right] c(x_0, t_0). \tag{6.23}$$

The only element on the right-hand side which depends upon x and t is the Gaussian function $\exp[-(x - x_0)^2/4D(t - t_0)]/\sqrt{4\pi D(t - t_0)}$. We have already seen that this function satisfies the diffusion equation. Provided that $c(x_0, t_0)$ is a well-behaved function of x_0, the operations of partial differentiation can be carried inside the integral, and it follows that equation (6.23) also satisfies the diffusion equation. Note that this solution was discussed earlier, where it occurred (with $t_0 = 0$) as equations (4.7) and (4.8). The difference is that here it was obtained directly from the random walk model of the microscopic particle motion, whereas earlier it was obtained by solving the diffusion equation using the convolution theorem.

At this point it will be useful to review this derivation of the diffusion equation. We started from the assumption that diffusing particles are following independent random walks. We used the central limit theorem to show that the probability density for the position of any one particle is a Gaussian function. We then used this to obtain the concentration $c(x, t)$ in the form of equation (6.23), and noted that this expression satisfies the diffusion equation, because the Gaussian factor is observed to satisfy the diffusion equation.

This derivation of the diffusion equation is valid, but a little unsatisfying in two respects. First, it does not show that random walks lead to the diffusion equation directly, but rather constructs an expression for the concentration which is observed to satisfy the diffusion equation. Secondly, this derivation is applicable only in an infinite medium, because the Gaussian solution is valid only in that case. The next section will discuss generalisations of the continuous random walk model for diffusing particles, which can take account of situations where the medium in which molecules diffuse is not homogeneous. It will lead to a generalised diffusion equation for the probability density, proceeding directly from the generalised random walk model. This alternative derivation will also be valid for diffusion in a finite medium.

6.4 The Fokker–Planck equation

The derivation of the diffusion equation in Section 6.3 started from the assumption that the particles undergo a continuous random walk, described by equation (6.7), with the statistics of the displacements specified by equation (6.8). Various extensions of this model arise in studying processes involving random motion. Further generalisations of the diffusion process will be introduced in Subsection 6.4.1. The probability density for such processes can be shown to satisfy a generalisation of the diffusion equation called the Fokker–Planck equation. This equation will be derived in Subsections 6.4.2 and 6.4.3.

6.4.1 Generalised diffusion processes

Consider a situation in which the particles are drifting with velocity v as well as making random steps. In this case the mean value of the displacement during the short time step δt is

$$\langle \Delta x_n \rangle = v \, \delta t. \tag{6.24}$$

We might also consider situations in which both the mean and variance of the displacement are functions of the current position of the particle. In general, we will assume that the displacement $\Delta x = x(t + \delta t) - x(t)$ of a particle is a random variable, with mean and variance given respectively by

$$\langle \Delta x \rangle = v(x(t), t) \, \delta t \tag{6.25}$$

and

$$\langle \Delta x^2 \rangle - \langle \Delta x \rangle^2 = 2D(x(t), t) \, \delta t. \tag{6.26}$$

Equations (6.25) and (6.26) will be taken as definitions of the drift velocity $v(x, t)$ and diffusion coefficient $D(x, t)$ in cases where these may depend upon position and/or time. (We emphasise that $v(x, t)$ in equation (6.25) is not the randomly varying velocity which causes diffusion, but an additional velocity which depends smoothly on time.) In the next two subsections we shall develop an extension of the diffusion equation for this more general type of random walk; this generalisation is known as the Fokker–Planck equation.

6.4.2 Probability density for generalised diffusion

The *generalised random walk* described by equations (6.25) and (6.26) can also be described by calculating the probability density $\rho(x,t)$ for the particle to reach position x at time t. In the following, we will derive a partial differential equation for $\rho(x,t)$, the *generalised diffusion equation* or *Fokker–Planck equation*. The derivation will be one of the hardest parts of this course, but the result, equation (6.41), is easily applied to obtain a partial differential equation for $\rho(x,t)$. These partial differential equations can often be solved by the methods discussed in Chapter 4.

The approach is to determine the probability density at time $t + \delta t$, and to obtain an expression for $\partial \rho/\partial t$:

$$\frac{\partial \rho}{\partial t} = \frac{\rho(x, t+\delta t) - \rho(x,t)}{\delta t} + O(\delta t). \tag{6.27}$$

This results in an expression for $\partial\rho/\partial t$ which is in the form of a generalisation of the diffusion equation.

Let the probability density be $\rho(x,t)$, so that the probability of a particle being located in a small interval between x and $x + \delta x$ at time t is $\delta P = \rho(x,t)\,\delta x$. At this stage, we assume that $\rho(x,t)$ is normalised, and that it may be differentiated as many times as are required for the subsequent calculations. At each time step, separated by δt, the particle jumps a random distance z. The displacement z is a random variable with probability density ρ_s. (Note that earlier we used the Δx to represent the random displacement, but it is convenient to change to using the single symbol z from here on.)

As before, the subscript 's' denotes 'step'.

This probability density may be a function of the position x of the particle before the jump and of time t, and is written $\rho_\mathrm{s}(z,x,t)$. Thus, the probability of a particle initially at position x at time t being found in a small interval between $x + z$ and $x + z + \delta z$ at time $t + \delta t$ is $\rho_\mathrm{s}(z,x,t)\,\delta z$.

The value of the displacement z at time t is independent of all of the previous steps. The probability of the particle reaching the interval $[x, x+\delta x]$ at time $t + \delta t$, having been in the interval $[x_0, x_0 + \delta x_0]$ at time t, is therefore the product of the probability for being in the first interval and the probability of jumping from the first interval to the second. This probability is

$$\delta \mathcal{P}(x_0) = \rho_\mathrm{s}(x - x_0, x_0, t)\,\delta x \times \rho(x_0, t)\,\delta x_0. \tag{6.28}$$

Let us now obtain an equation for the probability density at time $t + \delta t$, in terms of the probability density $\rho(x_0, t)$ at time t. The range of x_0 is divided into small intervals of width δx_0. The probability of reaching the interval $[x, x+\delta x]$ at time $t + \delta t$ is equal to the sum of the probabilities for reaching this interval from all of the intervals $[n\,\delta x_0, (n+1)\,\delta x_0]$. This is illustrated in Figure 6.4.

6.4 The Fokker–Planck equation

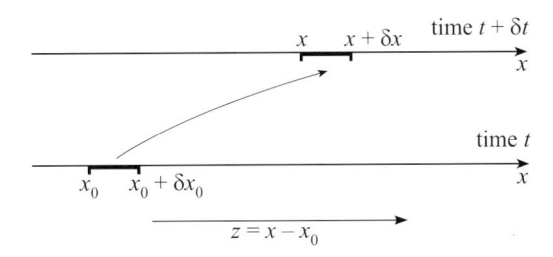

Figure 6.4 The probability to be in the interval $[x, x + \delta x]$ at time $t + \delta t$ is obtained by summing contributions from particles in all of the intervals $[x_0, x_0 + \delta x_0]$ at time t. The probability density for making a jump of $z = x - x_0$ from position x_0 at time t to position x at time $t + \delta t$ is $\rho_s(z, x_0, t)$.

Summing contributions in the form of equation (6.28), the probability of being in $[x, x + \delta x]$ at time $t + \delta t$ is

$$\rho(x, t + \delta t)\delta x = \delta x \sum_{n=-\infty}^{\infty} \delta x_0 \, \rho_s(x - n\, \delta x_0, x_0, t) \, \rho(n\, \delta x_0, t). \tag{6.29}$$

In the limit as $\delta x_0 \to 0$, the sum becomes an integral over x_0. Dividing both sides by δx and taking the limit as $\delta x \to 0$ and $\delta x_0 \to 0$ gives

$$\rho(x, t + \delta t) = \int_{-\infty}^{\infty} dx_0 \, \rho_s(x - x_0, x_0, t) \rho(x_0, t). \tag{6.30}$$

This equation and its derivation are analogous to that of equation (6.22), and also to the calculation in Section 5.3. In this case the difference between the initial and final times is small, but the distribution of the step $z = x - x_0$ need not be Gaussian, and may depend upon the initial position, x_0, of the particle.

6.4.3 Derivation of the Fokker–Planck equation

The objective is now to use equation (6.30) to determine a differential equation for the probability density $\rho(x, t)$; we expect that this will be some generalisation of the diffusion equation.

The function $\rho_s(z, x, t)$, considered as a function of z for any fixed value of x and t, is very sharply peaked about $z = 0$, reflecting the fact that the probability of making a long jump (significantly longer than $\sqrt{D\, \delta t}$) must be very small if the statistics of z satisfy $\langle z^2 \rangle = 2D\, \delta t$, as implied by equation (6.26). The probability density $\rho_s(x - x_0, x_0, t)$ for reaching x at time $t + \delta t$ starting from x_0 at time t, and the probability density $\rho(x, t)$ are illustrated schematically in Figure 6.5.

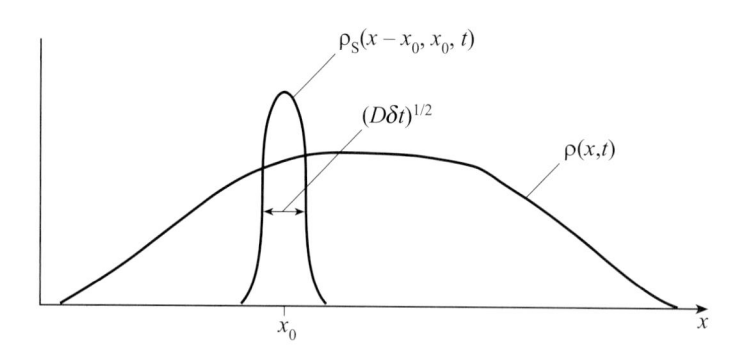

Figure 6.5 The probability density $\rho_s(x - x_0, x_0, t)$ for the step from x_0 at time t to x at $t + \delta t$ is very sharply peaked around x_0 (the width of the peak is approximately $\sqrt{D\,\delta t}$). The probability density $\rho(x, t)$ at time t is relatively slowly varying.

It is convenient to change the variable of integration in equation (6.30) to $z = x - x_0$, so that the dominant contribution to the integral comes from the region close to $z = 0$. Noting that $x_0 = x - z$, we have

$$\rho(x, t + \delta t) = \int_{-\infty}^{\infty} dz\, \rho_s(z, x - z, t)\rho(x - z, t). \tag{6.31}$$

We now consider how to approximate this expression, leading to a differential equation for $\rho(x, t)$. Given that the dominant contribution to the integral comes from the region around $z = 0$, we can approximate $\rho(x - z, t)$ by a Taylor series:

$$\rho(x - z, t) = \rho(x, t) - \frac{\partial \rho}{\partial x}(x, t)z + \tfrac{1}{2}\frac{\partial^2 \rho}{\partial x^2}(x, t)z^2 + \ldots$$

$$= \sum_{k=0}^{\infty} \frac{(-1)^k}{k!} \frac{\partial^k}{\partial x^k}\left[\rho(x, t)\right] z^k. \tag{6.32}$$

We also make a Taylor series expansion of $\rho_s(z, x - z, t)$ in its *second* argument only: we write

$$\rho_s(z, x - z, t) = \rho_s(z, x, t) - \frac{\partial \rho_s}{\partial x}(z, x, t)\, z$$

$$+ \tfrac{1}{2}\frac{\partial^2 \rho_s}{\partial x^2}(z, x, t)\, z^2 + O(z^3), \tag{6.33}$$

where it should be understood that the partial derivatives here are with respect to the second argument. Note that this expression is rather different in structure from equation (6.32) in that the coefficients of the Taylor series are themselves functions of z, which approach zero as $|z| \to \infty$. We can in fact combine both of these Taylor series together, and write

$$\rho_s(z, x - z, t)\rho(x - z, t) = \sum_{k=0}^{\infty} \frac{(-1)^k}{k!} \frac{\partial^k}{\partial x^k}\left[\rho_s(z, x, t)\rho(x, t)\right] z^k, \tag{6.34}$$

As indicated in Figure 6.5, close to $z = 0$, $\rho_s(z, x - z, t)$ is rapidly varying in its first argument but should be slowly varying in z from its second argument. For this reason, ρ_s is expanded in its second argument only.

which is a Taylor series for the function $\rho_s(z, x - \epsilon, t)\, \rho(x - \epsilon, t)$ expanded in powers of ϵ, which is then set to $\epsilon = z$. We now substitute this into equation (6.31), and obtain

$$\rho(x, t + \delta t) = \sum_{k=0}^{\infty} \frac{(-1)^k}{k!} \int_{-\infty}^{\infty} dz\, \frac{\partial^k}{\partial x^k}\left[\rho_s(z, x, t)\rho(x, t)\right] z^k$$

$$= \sum_{k=0}^{\infty} \frac{(-1)^k}{k!} \frac{\partial^k}{\partial x^k}\left[\rho(x, t)\int_{-\infty}^{\infty} dz\, \rho_s(z, x, t) z^k\right]. \tag{6.35}$$

In this second line we took everything that is independent of z outside the integral.

6.4 The Fokker–Planck equation

We can now define moments $M_k(x,t)$ of the probability density $\rho_s(z,x,t)$ of displacements z. By analogy with equation (1.31), these moments are defined by

$$M_k(x,t) = \int_{-\infty}^{\infty} dz\, z^k \rho_s(z,x,t). \tag{6.36}$$

With this definition, equation (6.35) becomes

$$\rho(x, t+\delta t) = \sum_{k=0}^{\infty} \frac{(-1)^k}{k!} \frac{\partial^k}{\partial x^k} \left[\rho(x,t) M_k(x,t) \right]. \tag{6.37}$$

The $k=0$ moment is unity because $\rho_s(z,x,t)$ is normalised. The first and second moments are obtained from equations (6.25) and (6.26), with Δx replaced by z. It follows from equation (6.26) that the magnitude of the typical displacement is $O(\delta t^{1/2})$. We will assume that all of the moments $M_k(x,t)$ are finite (that is, the integrals (6.36) are convergent), so that $\langle z^k \rangle = O(\delta t^{k/2})$ for $k > 2$. We therefore have

$$M_0(x,t) = 1$$
$$M_1(x,t) = v(x,t)\, \delta t$$
$$M_2(x,t) = 2D(x,t)\, \delta t + O(\delta t^2)$$
$$M_k(x,t) = O(\delta t^{k/2}),\ (k > 2). \tag{6.38}$$

Substituting these expressions into equation (6.37), and dropping terms in higher powers of the small quantity δt, we obtain

$$\rho(x, t+\delta t) = \rho(x,t) - \frac{\partial}{\partial x}\left[v(x,t)\,\rho(x,t)\right]\delta t$$
$$+ \tfrac{1}{2}\frac{\partial^2}{\partial x^2}\left[2D(x,t)\,\rho(x,t)\right]\delta t + O(\delta t^{3/2}). \tag{6.39}$$

From a Taylor expansion of $\rho(x,t)$, equation (6.27), we have

$$\rho(x,t+\delta t) = \rho(x,t) + \frac{\partial \rho}{\partial t}(x,t)\,\delta t + O(\delta t^2). \tag{6.40}$$

We now take the limit as $\delta t \to 0$. Equating the terms proportional to δt in equations (6.39) and (6.40) gives

$$\frac{\partial \rho}{\partial t} = -\frac{\partial}{\partial x}\left[v(x,t)\,\rho\right] + \frac{\partial^2}{\partial x^2}\left[D(x,t)\,\rho\right]. \tag{6.41}$$

This equation will be termed the *generalised diffusion equation*. In many texts it is called the *Fokker–Planck equation*. It reduces to the standard diffusion equation when $v=0$ and when D is independent of x and t.

The derivation above gives a direct route to obtaining the standard diffusion equation, as opposed to the indirect route discussed in Section 6.3, which constructs a solution and then observes that it satisfies the diffusion equation. The construction used in this subsection has the advantage that it is 'local', in the sense that $\rho(x,t+\delta t)$ only depends on $\rho(x',t)$ for values of x' which are close to x. The form of the resulting differential equation in the vicinity of x therefore depends on properties of the moments $M_k(x,t)$ at x, and is not affected by the boundaries of the system. The derivation given in this subsection is therefore valid for finite, as well as infinite, systems.

6.4.4 The Fokker–Planck equation in three dimensions

The calculation leading to the generalised diffusion equation was carried out in one space dimension. To understand the form of the generalised diffusion equation in three dimensions, we must first consider how the definition of the generalised random walk, discussed at the start of this section, extends to three dimensions. The continuous random walk in three dimensions was already introduced in Exercise 6.1 for the case where there is no drift and there is a constant diffusion coefficient D. Here we consider the case where there may be both drift and diffusion coefficients which depend upon both position and time. We also allow for the possibility that the diffusion process is not isotropic, that is, that the particles diffuse more easily in some directions than in others.

In three dimensions, the generalised random walk has a vector displacement $\Delta \boldsymbol{r}_n$ at the time step labelled by the integer index n. The components can be labelled by an index i, as in Subsection 6.2.2, so that the displacement may be written

$$\Delta \boldsymbol{r}_n = (\Delta x_n, \Delta y_n, \Delta z_n) = (\Delta x_{1,n}, \Delta x_{2,n}, \Delta x_{3,n}). \tag{6.42}$$

The mean displacements will, in general, be different in each direction, so that

$$\langle \Delta x_{i,n} \rangle = v_i(\boldsymbol{r}, t)\, \delta t, \tag{6.43}$$

where the drift velocity $\boldsymbol{v}(\boldsymbol{r}, t) = (v_1, v_2, v_3)$ can be a function of both position $\boldsymbol{r}$ and time t. We write the difference between the displacement $\Delta x_{i,n}$ and its mean value as $\delta x_{i,n} = \Delta x_{i,n} - v_i \delta t$. The fluctuating parts of the displacements $\delta x_{i,n}$ in different directions may be correlated. In general, we assume that they satisfy

$$\langle \delta x_{i,n}\, \delta x_{j,m} \rangle = 2\, D_{ij}(\boldsymbol{r}, t)\, \delta_{nm}\, \delta t. \tag{6.44}$$

The factor δ_{nm} is the Kronecker delta symbol, indicating that for different time steps ($n \neq m$) the displacements $\delta x_{i,n}$ have no correlation. The coefficients D_{ij} are elements of a 3×3 matrix which is termed the *diffusion matrix*.

This three-dimensional version of the generalised random walk can be described by giving the probability density for a particle to be at position $\boldsymbol{r}$ at time t. This probability density $\rho(\boldsymbol{r}, t)$ obeys a partial differential equation, which is a three-dimensional version of the generalised diffusion equation (6.41). The derivation follows exactly the same pattern as in the one-dimensional case, and we shall simply quote the result:

$$\frac{\partial \rho}{\partial t} = -\sum_{i=1}^{3} \frac{\partial}{\partial x_i}\left[v_i(\boldsymbol{r}, t)\, \rho\right] + \sum_{i=1}^{3}\sum_{j=1}^{3} \frac{\partial^2}{\partial x_i \partial x_j}\left[D_{ij}(\boldsymbol{r}, t)\, \rho\right]. \tag{6.45}$$

Exercise 6.3

What values of v_i and D_{ij} are appropriate for describing a diffusion process which is isotropic (occurs at the same rate in all directions), homogeneous (has the same properties at all points in space), time-independent, and which is without drift in three dimensions?

Show that substituting these values into equation (6.45) gives the usual three dimensional diffusion equation.

Exercise 6.4

Show that the generalised diffusion equations (6.41) and (6.45) are in the form of continuity equations.

What is the scalar flux density $J(x,t)$ in the one-dimensional case, and the vector flux density $\boldsymbol{J}(\boldsymbol{r},t)$ in the three-dimensional case?

The continuity equation was discussed in Chapter 3, where the one-dimensional and three-dimensional forms occur as equations (3.26) and (3.37) respectively.

6.5 Summary and outcomes

Equation (6.41) gives a partial differential equation defining the probability density for any generalised diffusion process in which particles undergo a random process that can be described by equations (6.25) and (6.26). The derivation was rather complicated, but all the hard work has been done, and you can now easily find a partial differential equation for the probability density of any generalised diffusion process.

After studying this chapter, you should:
- be aware of how the random walk can be generalised to higher dimensions and also generalised so that it becomes a continuous function of time;
- be aware that the central limit theorem implies that the propagator for diffusion in an infinite medium is a Gaussian function, and that this implies that the concentration obeys the diffusion equation;
- be aware of the Fokker–Planck equation as a description of generalised diffusion processes, and of the method used to derive it.

We conclude this chapter by giving some examples of generalised diffusion processes, introduced through a series of exercises in the final section.

6.6 Further Exercises

Exercise 6.5

Determine the form of the generalised diffusion equation in one dimension which applies when both D and v are constant. Show that

$$\rho(x,t) = \frac{1}{\sqrt{4\pi Dt}} \exp\left[\frac{-(x-x_0-vt)^2}{4Dt}\right] \qquad (6.46)$$

is a normalised solution, representing the probability density of a particle which is initially at position x_0 when $t=0$.

Exercise 6.6

Consider the motion of small particles dispersed in a liquid. If the particles are denser than the liquid, they will tend to sink to the bottom. If the particles are very small, such as the pollen grains used in experiments to demonstrate Brownian motion, this sinking effect is opposed by the random motion of the particles which results from molecules of the liquid colliding with the particles. The motion of the particles is described by the probability density $\rho(z,t)$ for a particle to have height z at time t.

The effects of gravity and of the viscosity of the fluid are modelled by assuming that the particles drift downwards at a constant velocity v. The effect of the random motion of the molecules is modelled by assuming that the particles have a diffusion constant D.

Write down the partial differential equation for the probability density of the diffusing particles. What is the *steady-state solution* of this equation, satisfying $\partial \rho / \partial t = 0$, in the case where the depth of the liquid is h?

Hint: Is this the same as the result from the previous exercise?

Exercise 6.7

Consider a situation where a particle (moving in one dimension) experiences a restoring force when displaced from the origin, such that a particle at position x moves with velocity $v = -\alpha x$ (where $\alpha > 0$). The particle also experiences random displacements which make it move diffusively, with diffusion constant D. Use equation (6.41) to show that the probability density for the particle satisfies

$$\frac{\partial \rho}{\partial t} = \alpha \frac{\partial}{\partial x}[x\,\rho] + D\frac{\partial^2 \rho}{\partial x^2}. \tag{6.47}$$

Solve this equation for a steady-state probability density in which the particle flux J is zero, in a region without boundaries.

Solutions to Exercises in Chapter 6

Solution 6.1

Here x, y and z are independent random variables. According to equation (1.20), the joint probability density is therefore a product: $\rho(x,y,z) = \rho_1(x)\,\rho_2(y)\,\rho_3(z)$, where ρ_1, ρ_2 and ρ_3 are probability densities for x, y and z, respectively. The statistics of the x-coordinate displacement specified in the question are exactly the same as for the one-dimensional continuous random walk treated in Subsection 6.2.1. Its probability density, for particles starting at $x = 0$, is given by equation (6.12), so

$$\rho_1(x) = \frac{1}{\sqrt{4\pi Dt}} \exp\left(\frac{-x^2}{4Dt}\right).$$

The statistics of y and z are the same as those of x; accordingly, the probability density at time t is

$$\rho(x,y,z) = \frac{1}{\sqrt{4\pi Dt}} \exp\left(\frac{-x^2}{4Dt}\right) \frac{1}{\sqrt{4\pi Dt}} \exp\left(\frac{-y^2}{4Dt}\right) \frac{1}{\sqrt{4\pi Dt}} \exp\left(\frac{-z^2}{4Dt}\right)$$

$$= \frac{1}{(4\pi Dt)^{3/2}} \exp\left[-\frac{(x^2+y^2+z^2)}{4Dt}\right].$$

This is analogous to the expression obtained in the two-dimensional case in Exercises 3.2 and 3.24, but here the result is obtained via a more direct route.

Solution 6.2

Let the typical time between collisions be δt, and let the typical speed of the molecules be $v = 260\,\mathrm{m\,s^{-1}}$. If we model the path of a molecule as a random walk, it is reasonable to suppose that the typical distance travelled between collisions is $\sigma = v\,\delta t$. Using equation (6.11), we have $\sigma^2 = v^2 \delta t^2 = 2D\,\delta t$, so $\delta t = 2D/v^2$. The typical distance between collisions is often referred to as the mean free path, and given the symbol λ. Our simple estimates give $\lambda \simeq v\,\delta t = 2D/v \simeq 10^{-7}\,\mathrm{m}$. Dividing by v we have $\delta t \simeq 4\times 10^{-10}\,\mathrm{s}$, so that the number of collisions per second is $N = 1/\delta t \simeq 2.6 \times 10^9$. Approximations based upon using the central limit theorem are therefore expected to be very precise, because N is so large.

Solution 6.3

If there is no drift, we set $\boldsymbol{v} = \boldsymbol{0}$. If the system is homogeneous and time-independent, the diffusion coefficients D_{ij} are independent of $\boldsymbol{r}$ and t. Because the diffusion is isotropic, the diffusion matrix must must not favour any one direction. Taking the diffusion matrix to be a multiple of the identity matrix satisfies this requirement, so that we set $D_{ij} = D\delta_{ij}$ (it can be shown that this is the only choice which makes the diffusion process isotropic). With these choices, equation (6.45) does indeed simplify to the standard three-dimensional diffusion equation:

$$\frac{\partial \rho}{\partial t} = D \sum_{i=1}^{3} \sum_{j=1}^{3} \delta_{ij} \frac{\partial^2 \rho}{\partial x_i \partial x_j}$$

$$= D \sum_{i=1}^{3} \frac{\partial^2 \rho}{\partial x_i^2} = D\nabla^2 \rho.$$

Solution 6.4

Equation (6.41) is in the form of a one-dimensional continuity equation, $\partial \rho/\partial t + \partial J/\partial x = 0$, with the flux J equal to

$$J = v\rho - \frac{\partial}{\partial x}(D\,\rho).$$

Equation (6.45) is also in the form of a three-dimensional continuity equation $\partial \rho/\partial t + \boldsymbol{\nabla} \cdot \boldsymbol{J} = 0$, with flux density $\boldsymbol{J}$ having components

$$J_i = v_i \rho - \sum_{j=1}^{3} \frac{\partial}{\partial x_j}\left[D_{ij}\,\rho\right].$$

Solution 6.5

Since v and D are constants, they can be taken outside the derivatives in equation (6.41). The generalised diffusion equation in this case is

$$\frac{\partial \rho}{\partial t} = -v\frac{\partial \rho}{\partial x} + D\frac{\partial^2 \rho}{\partial x^2}.$$

By differentiating the solution quoted in the exercise, it is seen to satisfy this differential equation. By integrating over x using a standard Gaussian integral, we see that this is a normalised solution, which can represent a probability density. By inspection, we see that this solution approaches zero in the limit as $t \to 0$ at all points except $x = x_0$, so the solution represents a motion in which the particle starts from position x_0.

Solution 6.6

The generalised diffusion equation is the same as for the previous exercise, except that the name of the space coordinate is z, and the sign of the velocity v is reversed, because positive v corresponds to decreasing z. The partial differential equation for ρ is

$$\frac{\partial \rho}{\partial t} = \frac{\partial}{\partial z}\left[v\rho + D\frac{\partial \rho}{\partial z}\right].$$

In the steady-state, ρ becomes independent of time so that $\partial \rho/\partial t = 0$. The quantity in square brackets is therefore independent of both z and t. Comparison with the continuity equation (3.26) shows that this quantity is minus the flux density, J. If the liquid is in a finite container, it may be assumed that the flux of particles across the upper surface of the liquid is zero. The zero-flux condition gives the following equation for $\rho(z)$:

$$0 = v\rho + D\frac{d\rho}{dz}.$$

This equation has the solution

$$\rho(z) = A\exp(-vz/D),$$

where A is a constant which is determined by normalising the distribution. If the depth of the liquid is h, the normalisation condition is

$$1 = \int_0^h dz\,\rho(z) = \frac{AD}{v}[1 - \exp(-vh/D)].$$

Rearranging this expression to find A, we obtain the required solution

$$\rho(z) = \frac{v\exp(-vz/D)}{D[1 - \exp(-vh/D)]}.$$

Solution 6.7

Substituting $v(x) = -\alpha x$ into equation (6.41) and letting D be constant gives the differential equation quoted in the exercise directly. This equation is in the form of a continuity equation, with flux density

$$J = -\alpha x \rho - D \frac{\partial \rho}{\partial x}.$$

When $\partial \rho/\partial t = 0$ and $J = 0$, this has a Gaussian solution. The solution of this differential equation was considered in Block I, Chapter 3, Exercise 3.25. The normalised probability density is then

$$\rho(x) = \frac{\exp(-\alpha x^2/2D)}{\sqrt{2\pi D/\alpha}}.$$

INDEX

approximate form for probability distribution   57
atoms   76
average   22
Avogadro's number   80

bias   47
binomial coefficient   55, 56
boundary conditions
    one-dimensional   122
Brown, Robert   50
Brownian motion   50

caloric model   96
central limit theorem   154, 155
coin tossing   14
coin-tossing function   44, 61
    statistics   45
combination of events   15
combined event   15
complementary error function   31
complex-valued function   165
concentration   78, 79, 86
    of heat energy   95
concentration of heat energy   95
continuity equation   87, 88
    one-dimensional   87
    three-dimensional   89
continuous random variable   17
    probability density   17
    probability density function   17
continuous random walk   179
convolution theorem   158
correlated random function   61
correlation   26
correlation coefficient   26, 46
    for discrete variables   26, 27
correlation function   46, 61

degenerate eigenvalues   133
die (dice) throwing
    moments   24
    probability   14–16
    statistics   23
Diffusion   76
diffusion   51, 77
    finite medium
      one dimension   121
      two and three dimensions   129
    generalised   185, 186
    infinite medium   116
    probability density   186
    semi-infinite medium   137
diffusion coefficient   52, 60
    for generalised diffusion   185
diffusion constant   75, 93
diffusion equation   60, 75, 93, 183
    generalised   189
    one-dimensional   75, 93
    three-dimensional   75, 94
diffusion matrix   190

diffusion–advection equation   101
Dirichlet boundary condition   136
discrete events   13
discrete random variable   21
Distribution, sums of random variables   158
divergence   90
doubly degenerate   133
drift   53, 54
drift velocity   53
    for generalised diffusion   185

eigenfunctions   123
    completeness   136
    normalised   128, 134
    orthogonality of   126
eigenvalues   123
Einstein, Albert   51
electricity   51
element of probability   17
elementary outcome   15
empirical value   14
energy   95
error function   30, 31
    complementary   31
event   13
    combined   15
    discrete   13
    mutually exclusive   15
expectation value   23
    for two variables   25
exponential distribution   19, 21, 24
exponential function   63

factorial   56
Fick's law   93
Fick, Adolf E.   93
flux   81, 83, 85, 86
flux density   78, 81, 86
    of heat energy   96
    one-dimensional   85
    scalar   82
    vector   83
Fokker–Planck equation   185, 189
    derivation   187
    in three dimensions   190
Fourier transform of a probability density   166
Fourier's law of heat conduction   96
frequency of an outcome   13
friction   77

gambler's fortune   51
gas   51
Gauss's theorem   91, 92
Gaussian approximation   163
    complex-valued function   165
Gaussian distribution   28, 60
Gaussian function   28, 30, 31, 63
Gaussian random variable   160
generalised diffusion   185
generalised diffusion equation   189

generalised Fourier series   134
generalised random walk   186
Gibbs, Josiah Willard   51

heat   96
heat equation   95, 96
Heaviside function   120
Helmholtz equation   130
    eigenvalues and eigenfunctions of   130
Helmholtz, Hermann von   130
histogram   28, 29
homogeneous   96
homogeneous system   78

independent events   16
independent random variables   21
induction, proof by   155, 171, 173
initial conditions   122
integer part function   58
isotropic   96

joint probability   21
joint probability density   20, 27

Kronecker delta symbol   46

Laplacian operator   75
linearly independent functions   133
Lorentzian function   163

Mach, Ernst   51
mass density   90
mass-specific heat capacity   95
mean value   23, 27
    of a normal distribution   30
mol   80
molar concentration   80
molecules   76
moments
    in throwing a die (dice)   24
    of a normal distribution   29
    of a random variable   24
mutually exclusive events   15

Neumann boundary condition   94, 130
normal distribution   28–32
    mean   30
    moments of   29
    probability density function   30
    standard deviation   30
    variance   30
normal distribution function   31
normal probability density function   30
normalisation condition   18
normalised function   18
normalised solution   98

orthogonality   127
orthogonality relation   124
    one-dimensional   128
    two- and three-dimensional   134
orthonormal   128
outcome   13
    elementary   15

parity   56
Pascal, Blaise   55
Pascal's triangle   55
Perrin, Jean   51
playing card probabilities   15, 16
Poisson distribution   34, 157, 170
price changes   51
probability   13–22
    in coin tossing   14
    in drawing playing cards   15, 16
    in successive trials   16
    in throwing a die (dice)   14–16
    of combined events   15
    of mutually exclusive events   15
probability density   17
    exponential distribution   19, 21, 24
    uniform distribution   18
probability density function
    joint   20, 27
    of a continuous random variable   17
    of a normal distribution   30
probability distribution
    approximate form   57
    of a random walk   54
propagator   117

random function   44
    with correlations   61, 62
random variable   22
    continuous   17
    discrete   21
    independent   21
    two or more   20, 25
random walk   48, 50, 180
    continuous   179
    generalised   186
    one-dimensional   179
    probability distribution   54
    simple   48, 49
    statistics   52
    three-dimensional   181
    two-dimensional   181, 182
    with drift   53, 54
rate constant   118
rate constant (Poisson distribution)   34
realisation   44
recurrence relation   55
relative error   58
running average   62

semi-infinite medium   115
separation constant   122
separation of variables   122
simple random walk   48, 49
sinc   163
special function   30
spectrum   130
standard deviation   24
    of a normal distribution   30
star motion   51
statistics   22–28

Index

    in throwing a die (dice)   23
    of a random walk   52
    of the coin-tossing function   45
steady flow   81, 90
steady state   90
step function   120
Stirling's formula   58
stochastic process   44
successive trials   16

Taylor series   60
temperature   95
temperature records   61
temperature waves   137
thermal conduction   96

thermal conductivity   96
thermal convection   77
thermal diffusivity   96
thermal insulation   121
thermal isolation   121
three-dimensional continuity equation   90
trial   13

uniform distribution   18

variance   24, 52
    of a normal distribution   30
vector flux density   83

weight function   63